bookdown

Authoring Books and Technical
Documents with R Markdown

Chapman & Hall/CRC
The R Series

Series Editors

John M. Chambers
Department of Statistics
Stanford University
Stanford, California, USA

Torsten Hothorn
Division of Biostatistics
University of Zurich
Switzerland

Duncan Temple Lang
Department of Statistics
University of California, Davis
Davis, California, USA

Hadley Wickham
RStudio
Boston, Massachusetts, USA

Aims and Scope

This book series reflects the recent rapid growth in the development and application of R, the programming language and software environment for statistical computing and graphics. R is now widely used in academic research, education, and industry. It is constantly growing, with new versions of the core software released regularly and more than 9,000 packages available. It is difficult for the documentation to keep pace with the expansion of the software, and this vital book series provides a forum for the publication of books covering many aspects of the development and application of R.

The scope of the series is wide, covering three main threads:

- Applications of R to specific disciplines such as biology, epidemiology, genetics, engineering, finance, and the social sciences.
- Using R for the study of topics of statistical methodology, such as linear and mixed modeling, time series, Bayesian methods, and missing data.
- The development of R, including programming, building packages, and graphics.

The books will appeal to programmers and developers of R software, as well as applied statisticians and data analysts in many fields. The books will feature detailed worked examples and R code fully integrated into the text, ensuring their usefulness to researchers, practitioners and students.

Published Titles

Stated Preference Methods Using R, *Hideo Aizaki, Tomoaki Nakatani, and Kazuo Sato*

Using R for Numerical Analysis in Science and Engineering, *Victor A. Bloomfield*

Event History Analysis with R, *Göran Broström*

Extending R, *John M. Chambers*

Computational Actuarial Science with R, *Arthur Charpentier*

Testing R Code, *Richard Cotton*

Statistical Computing in C++ and R, *Randall L. Eubank and Ana Kupresanin*

Basics of Matrix Algebra for Statistics with R, *Nick Fieller*

Reproducible Research with R and RStudio, Second Edition, *Christopher Gandrud*

R and MATLAB® *David E. Hiebeler*

Statistics in Toxicology Using R *Ludwig A. Hothorn*

Nonparametric Statistical Methods Using R, *John Kloke and Joseph McKean*

Displaying Time Series, Spatial, and Space-Time Data with R, *Oscar Perpiñán Lamigueiro*

Programming Graphical User Interfaces with R, *Michael F. Lawrence and John Verzani*

Analyzing Sensory Data with R, *Sébastien Lê and Theirry Worch*

Parallel Computing for Data Science: With Examples in R, C++ and CUDA, *Norman Matloff*

Analyzing Baseball Data with R, *Max Marchi and Jim Albert*

Growth Curve Analysis and Visualization Using R, *Daniel Mirman*

R Graphics, Second Edition, *Paul Murrell*

Introductory Fisheries Analyses with R, *Derek H. Ogle*

Data Science in R: A Case Studies Approach to Computational Reasoning and Problem Solving, *Deborah Nolan and Duncan Temple Lang*

Multiple Factor Analysis by Example Using R, *Jérôme Pagès*

bookdown
Authoring Books and Technical Documents with R Markdown

Yihui Xie

RStudio, Inc.

CRC Press
Taylor & Francis Group
Boca Raton   London   New York

CRC Press is an imprint of the
Taylor & Francis Group, an **informa** business

A CHAPMAN & HALL BOOK

CRC Press
Taylor & Francis Group
6000 Broken Sound Parkway NW, Suite 300
Boca Raton, FL 33487-2742

© 2017 by Taylor & Francis Group, LLC
CRC Press is an imprint of Taylor & Francis Group, an Informa business

No claim to original U.S. Government works

Printed on acid-free paper
Version Date: 20161121

International Standard Book Number-13: 978-1-138-70010-9 (Paperback)

Visit the Taylor & Francis Web site at
http://www.taylorandfrancis.com

and the CRC Press Web site at
http://www.crcpress.com

To Shao Yong (邵雍),
for sharing a secret joy with simple words;

月到天心处，风来水面时。
一般清意味，料得少人知。

and

To Hongzhi Zhengjue (宏智禅师),
for sharing the peace of an ending life with simple words.

梦幻空华，六十七年；
白鸟淹没，秋水连天。

Contents

List of Tables

List of Figures

Preface

This short book introduces an R package, **bookdown**, to change your workflow of writing books. It should be technically easy to write a book, visually pleasant to view the book, fun to interact with the book, convenient to navigate through the book, straightforward for readers to contribute or leave feedback to the book author(s), and more importantly, authors should not always be distracted by typesetting details.

The **bookdown** package is built on top of R Markdown (`http://rmarkdown.rstudio.com`), and inherits the simplicity of the Markdown syntax (you can learn the basics in five minutes; see Section 2.1), as well as the possibility of multiple types of output formats (PDF/HTML/Word/...). It has also added features like multi-page HTML output, numbering and cross-referencing figures/tables/sections/equations, inserting parts/appendices, and imported the GitBook style (`https://www.gitbook.com`) to create elegant and appealing HTML book pages. This book itself is an example of how you can produce a book from a series of R Markdown documents, and both the printed version and the online version can look professional. You can find more examples at `https://bookdown.org`.

Despite the package name containing the word "book", **bookdown** is not only for books. The "book" can be anything that consists of multiple R Markdown documents meant to be read in a linear sequence, such as course handouts, study notes, a software manual, a thesis, or even a diary. In fact, many **bookdown** features apply to single R Markdown documents as well (see Section 3.4).

Why read this book

Can we write a book in one source format, and generate the output to multiple formats? Traditionally books are often written with LaTeX or Microsoft Word. Either of these tools will make writing books a one-way trip and you cannot turn back: if you choose LaTeX, you typically end up only with a PDF document; if you work with Word, you are likely to have to stay in Word forever, and may also miss the many useful features and beautiful PDF output from LaTeX.

Can we focus on writing the content without worrying too much about typesetting? There seems a natural contradiction between content and appearance, and we always have to balance our time spent on these two aspects. No one can have a cake and eat it too, but it does not mean we cannot have a half and eat a half. We want our book to look reasonably pretty, and we also want to focus on the content. One possibility is to give up PDF temporarily, and what you may have in return is a pretty preview of your book as HTML web pages. LaTeX is an excellent typesetting tool, but you can be easily buried in the numerous LaTeX commands and typesetting details while you are working on the book. It is just so hard to refrain from previewing the book in PDF, and unfortunately also so common to find certain words exceed the page margin, certain figures float to a random page, five or six stray words at the very end of a chapter proudly take up a whole new page, and so on. If the book is to be printed, we will have to deal with these issues eventually, but it is not worth being distracted over and over again while you are writing book. The fact that the Markdown syntax is simpler and has fewer features than LaTeX also helps you focus on the content. Do you really have to define a new command like `\myprecious{}` that applies `\textbf{\textit{\textsf{}}}` to your text? Does the letter "R" have to be enclosed in `\proglang{}` when readers can easily figure out it stands for the R language? It does not make much difference whether everything, or nothing, needs the reader's attention.

Can readers interact with examples in our book as they read it? The answer is certainly no if the book is printed on paper, but it is possible if your book has an HTML version that contains live examples, such as Shiny applications (`https://shiny.rstudio.com`) or HTML widgets (`https://htmlwidgets.org`).

For example, readers may immediately know what happens if they change certain parameters of a statistical model.

Can we get feedback and even contributions from readers as we develop the book? Traditionally the editor will find a small number of anonymous reviewers to review your book. Reviewers are often helpful, but you may still miss the wisdom of more representative readers. It is too late after the first edition is printed, and readers may need to wait for a few years before the second edition is ready. There are some web platforms that make it easy for people to provide feedback and contribute to your projects. GitHub (https://github.com) is one prominent example. If anyone finds a typo in your book, he/she can simply correct it online and submit the change back to you for your approval. It is a matter of clicking a button to merge the change, with no questions asked or emails back and forth. To be able to use these platforms, you need to learn the basics of version control tools like GIT, and your book source files should be in plain text.

The combination of R (https://www.r-project.org), Markdown, and Pandoc (http://pandoc.org) makes it possible to go from one simple source format (R Markdown) to multiple possible output formats (PDF, HTML, EPUB, and Word, etc.). The **bookdown** package is based on R Markdown, and provides output formats for books and long-form articles, including the Git-Book format, which is a multi-page HTML output format with a useful and beautiful user interface. It is much easier to typeset in HTML than LaTeX, so you can always preview your book in HTML, and work on PDF after the content is mostly done. Live examples can be easily embedded in HTML, which can make the book more attractive and useful. R Markdown is a plain-text format, so you can also enjoy the benefits of version control, such as collaborating on GitHub. We have also tried hard to port some important features from LaTeX to HTML and other output formats, such as figure/table numbering and cross-references.

In short, you just prepare a few R Markdown book chapters, and **bookdown** can help you turn them into a beautiful book.

Structure of the book

Chapters 1 and 2 introduce the basic usage and syntax, which should be sufficient to get most readers started in writing a book. Chapters 3 and 4 are for those who want to fine-tune the appearance of their books. They may look very technical if you are not familiar with HTML/CSS and LaTeX. You do not need to read these two chapters very carefully for the first time. You can learn what can be possibly changed, and come back later to know how. For Chapter 5, the technical details are not important unless you do not use the RStudio IDE (Section 5.4). Similarly, you may feel overwhelmed by the commands presented in Chapter 6 to publish your book, but again, we have tried to make it easy to publish your book online via the RStudio IDE. The custom commands and functions are only for those who choose not to use RStudio's service or want to understand the technical details.

To sum it up, this book is a comprehensive reference of the **bookdown** package. You can follow the 80/20 rule when reading it. Some sections are there for the sake of completeness, and not all sections are equally useful to the particular book(s) that you intend to write.

Software information and conventions

This book is primarily about the R package **bookdown**, so you need to at least install R and the **bookdown** package. However, your book does not have to be related to the R language at all. It can use other computing languages (C++, SQL, Python, and so on; see Appendix B), and it can even be totally irrelevant to computing (e.g., you can write a novel, or a collection of poems). The software tools required to build a book are introduced in Appendix A.

The R session information when compiling this book is shown below:

```
sessionInfo()
```

```
## R version 3.3.2 (2016-10-31)
## Platform: x86_64-apple-darwin13.4.0 (64-bit)
## Running under: macOS Sierra 10.12.1
##
## locale:
## [1] en_US.UTF-8/en_US.UTF-8/en_US.UTF-8/C/en_US.UTF-8/en_US.UTF-8
##
## attached base packages:
## [1] stats     graphics  grDevices utils     datasets
## [6] base
##
## loaded via a namespace (and not attached):
## [1] bookdown_0.2    miniUI_0.1.1    rmarkdown_1.1
## [4] tools_3.3.2     shiny_0.14.2    htmltools_0.3.6
## [7] knitr_1.15
```

We do not add prompts (> and +) to R source code in this book, and we comment out the text output with two hashes ## by default, as you can see from the R session information above. This is for your convenience when you want to copy and run the code (the text output will be ignored since it is commented out). Package names are in bold text (e.g., **rmarkdown**), and inline code and filenames are formatted in a typewriter font (e.g., `knitr::knit('foo.Rmd')`). Function names are followed by parentheses (e.g., `bookdown::render_book()`). The double-colon operator `::` means accessing an object from a package.

Acknowledgments

First I'd like to thank my employer, RStudio, for providing me the opportunity to work on this exciting project. I was hoping to work on it when I first saw the GitBook project in 2013, because I immediately realized it was a beautiful book style and there was a lot more power we could add to it, judging from my experience of writing the **knitr** book (Xie, 2015) and reading other books. R Markdown became mature after two years, and luckily, **bookdown** became my official job in late 2015. There are not many things in the world better than the fact that your job happens to be your hobby (or vice versa). I totally enjoyed messing around JavaScript libraries, LaTeX packages, and endless regular expressions in R. Honestly I should also thank StackOverflow (http://stackoverflow.com), and I believe you all know what I mean,[1] if you have ever written any program code.

This project is certainly not a single person's effort. Several colleagues at RStudio have helped me along the way. Hadley Wickham provided a huge amount of feedback during the development of **bookdown**, as he was working on his book *R for Data Science* with Garrett Grolemund. JJ Allaire and Jonathan McPherson provided a lot of technical help directly to this package as well as support in the RStudio IDE. Jeff Allen, Chaita Chaudhari, and the RStudio Connect team have been maintaining the https://bookdown.org website. Robby Shaver designed a nice cover image for this book. Both Hadley Wickham and Mine Cetinkaya-Rundel reviewed the manuscript and gave me a lot of helpful comments. Tareef Kawaf tried his best to help me become a professional software engineer. It is such a blessing to work in this company with enthusiastic and smart people. I remember once I told Jonathan, "hey I found a problem in caching HTML widgets dependencies and finally figured out a possible solution". Jonathan grabbed his beer and said, "I already solved it." "Oh, nice, nice."

I also received a lot of feedback from book authors outside RStudio, including Jan de Leeuw, Jenny Bryan, Dean Attali, Rafael Irizarry, Michael Love, Roger Peng, Andrew Clark, and so on. Some users also contributed code to the project and helped revise the book. Here is a list of all contributors: https://github.com/rstudio/bookdown/graphs/contributors. It feels good when you invent a tool and realize you are also the beneficiary of your own tool. As someone who loves the GitHub pull request model, I wished readers did not have to email me there was a typo or obvious mistake in my book, but could just fix it via a pull request. This was made possible in **bookdown**. You can see how many pull requests on typos I have merged: https://github.com/rstudio/bookdown/pulls. It is nice to have so many outsourced careful human spell checkers. It is not that I do not know how

---

[1] http://bit.ly/2cWbiAp

to use a real spell checker, but I do not want to do this before the book is finished, and the evil Yihui also wants to leave a few simple tasks to the readers to engage them in improving the book.

The **bookdown** package is not possible without a few open-source software packages. In particular, Pandoc, GitBook, jQuery, and the dependent R packages, not to mention R itself. I thank the developers of these packages.

I moved to Omaha, Nebraska, in 2015, and enjoyed one year at Steeplechase Apartments, where I lived comfortably while developing the **bookdown** package, thanks to the extremely friendly and helpful staff. Then I met a professional and smart realtor, Kevin Schaben, who found a fabulous home for us in an amazingly short period of time, and I finished this book in our new home.

John Kimmel, the editor from Chapman & Hall/CRC, helped me publish my first book. It is my pleasure to work with him again. He generously agreed to let me keep the online version of this book for free, so I can continue to update it after it is printed and published (i.e., you do not have to wait for years for the second edition to correct mistakes and introduce new features). I wish I could be as open-minded as he is when I'm his age. Rebecca Condit and Suzanne Lassandro proofread the manuscript, and their suggestions were professional and helpful. Shashi Kumar solved some of my technical issues with the publisher's LaTeX class (`krantz.cls`) when I was trying to integrate it with **bookdown**. I also appreciate the very helpful comments from the reviewers Jan de Leeuw, Karl Broman, Brooke Anderson, Michael Grayling, Daniel Kaplan, and Max Kuhn.

Lastly I want to thank my family, in particular, my wife and son, for their support. The one-year-old has discovered that my monitor will light up when he touches my keyboard, so occasionally he just creeps into my office and presses randomly on the keyboard when I'm away. I'm not sure if this counts as his contribution to the book... @)!%)&@*

Yihui Xie
Elkhorn, Nebraska

About the Author

Yihui Xie (http://yihui.name) is a software engineer at RStudio (http://www.rstudio.com). He earned his PhD from the Department of Statistics, Iowa State University. He is interested in interactive statistical graphics and statistical computing. As an active R user, he has authored several R packages, such as **knitr**, **bookdown**, **animation**, **DT**, **tufte**, **formatR**, **fun**, **mime**, **highr**, **servr**, and **Rd2roxygen**, among which the **animation** package won the 2009 John M. Chambers Statistical Software Award (ASA). He also co-authored a few other R packages, including **shiny**, **rmarkdown**, and **leaflet**.

In 2006, he founded the Capital of Statistics (http://cos.name), which has grown into a large online community on statistics in China. He initiated the Chinese R conference in 2008, and has been involved in organizing R conferences in China since then. During his PhD training at Iowa State University, he won the Vince Sposito Statistical Computing Award (2011) and the Snedecor Award (2012) in the Department of Statistics.

He occasionally rants on Twitter (https://twitter.com/xieyihui), and most of the time you can find him on GitHub (https://github.com/yihui).

He enjoys spicy food as much as classical Chinese literature.

1

Introduction

This book is a guide to authoring books and technical documents with R Markdown (Allaire et al., 2016a) and the R package **bookdown** (Xie, 2016a). It focuses on the features specific to writing books, long-form articles, or reports, such as:

- how to typeset equations, theorems, figures and tables, and cross-reference them;
- how to generate multiple output formats such as HTML, PDF, and e-books for a single book;
- how to customize the book templates and style different elements in a book;
- editor support (in particular, the RStudio IDE); and
- how to publish a book.

It is not a comprehensive introduction to R Markdown or the **knitr** package (Xie, 2016c), on top of which **bookdown** was built. To learn more about R Markdown, please check out the online documentation http://rmarkdown. rstudio.com. For **knitr**, please see Xie (2015). You do not have to be an expert of the R language (R Core Team, 2016) to read this book, but you are expected to have some basic knowledge about R Markdown and **knitr**. For beginners, you may get started with the cheatsheets at https://www.rstudio. com/resources/cheatsheets/. The appendix of this book contains brief introductions to these software packages. To be able to customize the book templates and themes, you should be familiar with LaTeX, HTML and CSS.

1.1 Motivation

Markdown is a wonderful language to write relatively simple documents that contain elements like sections, paragraphs, lists, links, and images, etc. Pandoc (http://pandoc.org) has greatly extended the original Markdown syntax,[1] and added quite a few useful new features, such as footnotes, citations, and tables. More importantly, Pandoc makes it possible to generate output documents of a large variety of formats from Markdown, including HTML, LaTeX/PDF, Word, and slides.

There are still a few useful features missing in Pandoc's Markdown at the moment that are necessary to write a relatively complicated document like a book, such as automatic numbering of figures and tables in the HTML output, cross-references of figures and tables, and fine control of the appearance of figures (e.g., currently it is impossible to specify the alignment of images using the Markdown syntax). These are some of the problems that we have addressed in the **bookdown** package.

Under the constraint that we want to produce the book in multiple output formats, it is nearly impossible to cover all possible features specific to these diverse output formats. For example, it may be difficult to reinvent a certain complicated LaTeX environment in the HTML output using the (R) Markdown syntax. Our main goal is not to replace *everything* with Markdown, but to cover *most* common functionalities required to write a relatively complicated document, and make the syntax of such functionalities consistent across all output formats, so that you only need to learn one thing and it works for all output formats.

Another goal of this project is to make it easy to produce books that look visually pleasant. Some nice existing examples include GitBook (https://www.gitbook.com), Tufte CSS (http://edwardtufte.github.io/tufte-css/), and Tufte-LaTeX (https://tufte-latex.github.io/tufte-latex/). We hope to integrate these themes and styles into **bookdown**, so authors do not have to dive into the details of how to use a certain LaTeX class or how to configure CSS for HTML output.

---

[1]http://daringfireball.net/projects/markdown/

1.2 Get started

The easiest way for beginners to get started with writing a book with R Markdown and **bookdown** is through the demo `bookdown-demo` on GitHub:

1. Download the GitHub repository `https://github.com/rstudio/bookdown-demo` as a Zip file,[2] then unzip it locally.

2. Install the RStudio IDE. Note that you need a version higher than 1.0.0. Please download the latest version[3] if your RStudio version is lower than 1.0.0.

3. Install the R package **bookdown**:

    ```
    # stable version on CRAN
    install.packages("bookdown")
    # or development version on GitHub
    # devtools::install_github('rstudio/bookdown')
    ```

4. Open the `bookdown-demo` repository you downloaded in RStudio by clicking `bookdown-demo.Rproj`.

5. Open the R Markdown file `index.Rmd` and click the button `Build Book` on the `Build` tab of RStudio.

Now you should see the index page of this book demo in the RStudio Viewer. You may add or change the R Markdown files, and hit the `Knit` button again to preview the book. If you prefer not to use RStudio, you may also compile the book through the command line. See the next section for details.

Although you see quite a few files in the `bookdown-demo` example, most of them are not essential to a book. If you feel overwhelmed by the number of files, you can use this minimal example instead, which is essentially one file `index.Rmd`: `https://github.com/yihui/bookdown-minimal`. The

---

[2]`https://github.com/rstudio/bookdown-demo/archive/master.zip`
[3]`https://www.rstudio.com/products/rstudio/download/`

bookdown-demo example contains some advanced settings that you may want to learn later, such as how to customize the LaTeX preamble, tweak the CSS, and build the book on GitHub, etc.

1.3 Usage

A typical **bookdown** book contains multiple chapters, and one chapter lives in one R Markdown file, with the filename extension .Rmd. Each R Markdown file must start immediately with the chapter title using the first-level heading, e.g., # Chapter Title. All R Markdown files must be encoded in UTF-8, especially when they contain multi-byte characters such as Chinese, Japanese, and Korean. Here is an example (the bullets are the filenames, followed by the file content):

- index.Rmd

```
# Preface {-}

In this book, we will introduce an interesting
method.
```

- 01-intro.Rmd

```
# Introduction

This chapter is an overview of the methods that
we propose to solve an **important problem**.
```

- 02-literature.Rmd

```
# Literature

Here is a review of existing methods.
```

- 03-method.Rmd

  ```
  # Methods

  We describe our methods in this chapter.
  ```

- 04-application.Rmd

  ```
  # Applications

  Some _significant_ applications are demonstrated
  in this chapter.

  ## Example one

  ## Example two
  ```

- 05-summary.Rmd

  ```
  # Final Words

  We have finished a nice book.
  ```

By default, **bookdown** merges all Rmd files by the order of filenames, e.g., 01-intro.Rmd will appear before 02-literature.Rmd. Filenames that start with an underscore _ are skipped. If there exists an Rmd file named index.Rmd, it will always be treated as the first file when merging all Rmd files. The reason for this special treatment is that the HTML file index.html to be generated from index.Rmd is usually the default index file when you view a website, e.g., you are actually browsing http://yihui.name/index.html when you open http://yihui.name/.

You can override the above behavior by including a configuration file named _bookdown.yml in the book directory. It is a YAML file (https://en.wikipedia.org/wiki/YAML), and R Markdown users should be familiar with this format since it is also used to write the metadata in the beginning of R Markdown documents (you can learn more about YAML in Section B.2). You can use a field named `rmd_files` to define your own list and order of Rmd files for the book. For example,

```
rmd_files: ["index.Rmd", "abstract.Rmd", "intro.Rmd"]
```

In this case, **bookdown** will just use whatever you defined in this YAML field without any special treatments of `index.Rmd` or underscores. If you want both HTML and LaTeX/PDF output from the book, and use different Rmd files for HTML and LaTeX output, you may specify these files for the two output formats separately, e.g.,

```
rmd_files:
  html: ["index.Rmd", "abstract.Rmd", "intro.Rmd"]
  latex: ["abstract.Rmd", "intro.Rmd"]
```

Although we have been talking about R Markdown files, the chapter files do not actually have to be R Markdown. They can be plain Markdown files (`.md`), and do not have to contain R code chunks at all. You can certainly use **bookdown** to compose novels or poems!

At the moment, the major output formats that you may use include `bookdown::pdf_book`, `bookdown::gitbook`, `bookdown::html_book`, and `bookdown::epub_book`. There is a `bookdown::render_book()` function similar to `rmarkdown::render()`, but it was designed to render *multiple* Rmd documents into a book using the output format functions. You may either call this function from command line directly, or click the relevant buttons in the RStudio IDE. Here are some command-line examples:

```
bookdown::render_book("foo.Rmd", "bookdown::gitbook")
bookdown::render_book("foo.Rmd", "bookdown::pdf_book")
bookdown::render_book("foo.Rmd", bookdown::gitbook(lib_dir = "libs"))
bookdown::render_book("foo.Rmd", bookdown::pdf_book(keep_tex = TRUE))
```

To use `render_book` and the output format functions in the RStudio IDE, you can define a YAML field named `site` that takes the value bookdown::bookdown_site,[4] and the output format functions can be used in the output field, e.g.,

```
---
site: "bookdown::bookdown_site"
output:
  bookdown::gitbook:
    lib_dir: "book_assets"
  bookdown::pdf_book:
    keep_tex: yes
---
```

Then you can click the Build Book button in the Build pane in RStudio to compile the Rmd files into a book, or click the Knit button on the toolbar to preview the current chapter.

More **bookdown** configuration options in _bookdown.yml are explained in Section 4.4. Besides these configurations, you can also specify some Pandoc-related configurations in the YAML metadata of the *first* Rmd file of the book, such as the title, author, and date of the book, etc. For example:

```
---
title: "Authoring A Book with R Markdown"
author: "Yihui Xie"
date: "`r Sys.Date()`"
site: "bookdown::bookdown_site"
output:
  bookdown::gitbook: default
documentclass: book
bibliography: ["book.bib", "packages.bib"]
biblio-style: apalike
link-citations: yes
---
```

---

[4]This function calls bookdown::render_book().

1.4 Two rendering approaches

Merging all chapters into one Rmd file and knitting it is one way to render the book in **bookdown**. There is actually another way: you may knit each chapter in a *separate* R session, and **bookdown** will merge the Markdown output of all chapters to render the book. We call these two approaches "Merge and Knit" (M-K) and "Knit and Merge" (K-M), respectively. The differences between them may seem subtle, but can be fairly important depending on your use cases.

- The most significant difference is that M-K runs *all* code chunks in all chapters in the same R session, whereas K-M uses separate R sessions for individual chapters. For M-K, the state of the R session from previous chapters is carried over to later chapters (e.g., objects created in previous chapters are available to later chapters, unless you deliberately deleted them); for K-M, all chapters are isolated from each other.[5] If you want each chapter to compile from a clean state, use the K-M approach. It can be very tricky and difficult to restore a running R session to a completely clean state if you use the M-K approach. For example, even you detach/unload packages loaded in a previous chapter, R will not clean up the S3 methods registered by these packages.
- Because **knitr** does not allow duplicate chunk labels in a source document, you need to make sure there are no duplicate labels in your book chapters when you use the M-K approach, otherwise **knitr** will signal an error when knitting the merged Rmd file. Note that this means there must not be duplicate labels throughout the whole book. The K-M approach only requires no duplicate labels within any single Rmd file.
- K-M does not allow Rmd files to be in subdirectories, but M-K does.

The default approach in **bookdown** is M-K. To switch to K-M, you either use the argument new_session = TRUE when calling render_book(), or set new_session: yes in the configuration file _bookdown.yml.

You can configure the book_filename option in _bookdown.yml for the K-M

---

[5]Of course, no one can stop you from writing out some files in one chapter, and reading them in another chapter. It is hard to isolate these kinds of side-effects.

approach, but it should be a Markdown filename, e.g., `_main.md`, although the filename extension does not really matter, and you can even leave out the extension, e.g., just set `book_filename: _main`. All other configurations work for both M-K and K-M.

1.5 Some tips

Typesetting under the paging constraint (e.g., for LaTeX/PDF output) can be an extremely tedious and time-consuming job. I'd recommend you not to look at your PDF output frequently, since most of the time you are very unlikely to be satisfied: text may overflow into the page margin, figures may float too far away, and so on. Do not try to make things look right *immediately*, because you may be disappointed over and over again as you keep on revising the book, and things may be messed up again even if you only made some minor changes (see `http://bit.ly/tbrLtx` for a nice illustration).

If you want to preview the book, preview the HTML output. Work on the PDF version after you have finished the content of the book, and are very sure no major revisions will be required.

If certain code chunks in your R Markdown documents are time-consuming to run, you may cache them by adding the chunk option `cache = TRUE` in the chunk header, and you are recommended to label such code chunks as well, e.g.,

```
```{r important-computing, cache=TRUE}
```

In Chapter 5, we will talk about how to quickly preview a book as you edit. In short, you can use the `preview_chapter()` function to render a single chapter instead of the whole book. The function `serve_book()` makes it easy to live-preview HTML book pages: whenever you modify an Rmd file, the book can be recompiled and the browser can be automatically refreshed accordingly.
```

2

Components

In this chapter, we show the syntax of common components of a book, including R code, figures, tables, citations, math theorems, and equations, etc. First we start with the syntax of Pandoc's Markdown.

2.1 Markdown syntax

In this section We give a very brief introduction to Pandoc's Markdown. Readers who are familiar with Markdown can skip this section. The comprehensive syntax of Pandoc's Markdown can be found on the Pandoc website http://pandoc.org.

2.1.1 Inline formatting

You can make text *italic* by surrounding it with underscores or asterisks, e.g., _text_ or *text*. For **bold** text, use two underscores (__text__) or asterisks (**text**). Text surrounded by ~ will be converted to a subscript (e.g., H~2~SO~4~ renders H_2SO_4), and similarly, two carets (^) produce a superscript (e.g., Cl0^-1^ renders ClO^{-1}). To mark text as inline code, use a pair of backticks, e.g., `code`.[1] Small caps can be produced by the HTML tag span, e.g., Small Caps renders Small Caps. Links are created using [text](link), e.g., [RStudio](https://www.rstudio.com), and the syntax for images is similar: just add an exclamation mark, e.g., ![alt text or image title](path/to/image). Footnotes are put inside the square brackets after

---

[1] To include literal backticks, use more backticks outside, e.g., you can use two backticks to preserve one backtick inside: `` `code` ``.

a caret ˆ[], e.g., ˆ[This is a footnote.]. We will talk about citations in
Section 2.8.

2.1.2   Block-level elements

Section headers can be written after a number of pound signs, e.g.,

```
# First-level header
```

```
## Second-level header
```

```
### Third-level header
```

If you do not want a certain heading to be numbered, you can add {-} after
the heading, e.g.,

```
# Preface {-}
```

Unordered list items start with *, -, or +, and you can nest one list within
another list by indenting the sub-list by four spaces, e.g.,

```
- one item
- one item
- one item
    - one item
    - one item
```

The output is:

- one item
- one item
- one item
    - one item
    - one item

Ordered list items start with numbers (the rule for nested lists is the same
as above), e.g.,

```
1. the first item
2. the second item
3. the third item
```

The output does not look too much different with the Markdown source:

1. the first item
2. the second item
3. the third item

Blockquotes are written after >, e.g.,

```
> "I thoroughly disapprove of duels. If a man should challenge me,
  I would take him kindly and forgivingly by the hand and lead him
  to a quiet place and kill him."
>
> --- Mark Twain
```

The actual output (we customized the style for blockquotes in this book):

"I thoroughly disapprove of duels. If a man should challenge me, I would take him kindly and forgivingly by the hand and lead him to a quiet place and kill him."

— Mark Twain

Plain code blocks can be written after three or more backticks, and you can also indent the blocks by four spaces, e.g.,

```
```

This text is displayed verbatim / preformatted
```
```

Or indent by four spaces:

    This text is displayed verbatim / preformatted

2.1.3 Math expressions

Inline LaTeX equations can be written in a pair of dollar signs using the
LaTeX syntax, e.g., `$f(k) = {n \choose k} p^{k} (1-p)^{n-k}$` (actual output: $f(k) = \binom{n}{k} p^k (1-p)^{n-k}$); math expressions of the display style can be
written in a pair of double dollar signs, e.g., `$$f(k) = {n \choose k} p^{k}`
`(1-p)^{n-k}$$`, and the output looks like this:

$$f\left(k\right) = \binom{n}{k} p^k \left(1 - p\right)^{n-k}$$

You can also use math environments inside `$ $` or `$$ $$`, e.g.,

```
$$\begin{array}{ccc}
x_{11} & x_{12} & x_{13}\\
x_{21} & x_{22} & x_{23}
\end{array}$$
```

$$\begin{array}{ccc}
x_{11} & x_{12} & x_{13}\\
x_{21} & x_{22} & x_{23}
\end{array}$$

```
$$X = \begin{bmatrix}1 & x_{1}\\
1 & x_{2}\\
1 & x_{3}
\end{bmatrix}$$
```

$$X = \begin{bmatrix}1 & x_1\\ 1 & x_2\\ 1 & x_3\end{bmatrix}$$

```
$$\begin{vmatrix}a & b\\
c & d
\end{vmatrix}=ad-bc$$
```

$$\begin{vmatrix} a & b \\ c & d \end{vmatrix} = ad - bc$$

2.2 Markdown extensions by bookdown

Although Pandoc's Markdown is much richer than the original Markdown syntax, it still has a number of things that we may need for academic writing. For example, it supports math equations, but you cannot number and reference equations in multi-page HTML or EPUB output. We have provided a few Markdown extensions in **bookdown** to fill the gaps.

2.2.1 Number and reference equations

To number and refer to equations, put them in the equation environments and assign labels to them using the syntax (`\#eq:label`), e.g.,

```
\begin{equation}
  f\left(k\right) = \binom{n}{k} p^k\left(1-p\right)^{n-k}
  (\#eq:binom)
\end{equation}
```

It renders the equation below:

$$f(k) = \binom{n}{k} p^k (1-p)^{n-k} \tag{2.1}$$

You may refer to it using `\@ref(eq:binom)`, e.g., see Equation (2.1).

Equation labels must start with the prefix eq: in **bookdown**. All labels in **bookdown** must only contain alphanumeric characters, :, -, and/or /. Equation references work best for LaTeX/PDF output, and they are not well supported in Word output or e-books. For HTML output, **bookdown** can only number the equations with labels. Please make sure equations without labels are not numbered by either using the equation* environment or adding \nonumber or \notag to your equations. The same rules apply to other math environments, such as eqnarray, gather, align, and so on (e.g., you can use the align* environment).

We demonstrate a few more math equation environments below. Here is an unnumbered equation using the equation* environment:

```
\begin{equation*}
\frac{d}{dx}\left( \int_{a}^{x} f(u)\,du\right)=f(x)
\end{equation*}
```

$$\frac{d}{dx} \left(\int_a^x f(u)\, du \right) = f(x)$$

Below is an align environment (2.2):

```
\begin{align}
g(X_{n}) &= g(\theta)+g'({\tilde{\theta}})(X_{n}-\theta) \notag \\
\sqrt{n}[g(X_{n})-g(\theta)] &= g'\left({\tilde{\theta}}\right)
   \sqrt{n}[X_{n}-\theta ] (\#eq:align)
\end{align}
```

$$g(X_n) = g(\theta) + g'(\tilde{\theta})(X_n - \theta)$$
$$\sqrt{n}[g(X_n) - g(\theta)] = g'\left(\tilde{\theta}\right)\sqrt{n}[X_n - \theta] \tag{2.2}$$

You can use the split environment inside equation so that all lines share the same number (2.3). By default, each line in the align environment will be assigned an equation number. We suppressed the number of the first

line in the previous example using \notag. In this example, the whole split
environment was assigned a single number.

```
\begin{equation}
\begin{split}
\mathrm{Var}(\hat{\beta}) & =\mathrm{Var}((X'X)^{-1}X'y)\\
 & =(X'X)^{-1}X'\mathrm{Var}(y)((X'X)^{-1}X')'\\
 & =(X'X)^{-1}X'\mathrm{Var}(y)X(X'X)^{-1}\\
 & =(X'X)^{-1}X'\sigma^{2}IX(X'X)^{-1}\\
 & =(X'X)^{-1}\sigma^{2}
\end{split}
(\#eq:var-beta)
\end{equation}
```

$$
\begin{aligned}
\mathrm{Var}(\hat{\beta}) &= \mathrm{Var}((X'X)^{-1}X'y)\\
&= (X'X)^{-1}X'\mathrm{Var}(y)((X'X)^{-1}X')'\\
&= (X'X)^{-1}X'\mathrm{Var}(y)X(X'X)^{-1}\\
&= (X'X)^{-1}X'\sigma^2 IX(X'X)^{-1}\\
&= (X'X)^{-1}\sigma^2
\end{aligned} \tag{2.3}
$$

2.2.2 Theorems and proofs

Theorems and proofs are commonly used in articles and books in mathematics. However, please do not be misled by the names: a "theorem" is just a
numbered/labeled environment, and it does not have to be a mathematical
theorem (e.g., it can be an example irrelevant to mathematics). Similarly, a
"proof" is an unnumbered environment. In this section, we always use the
general meanings of a "theorem" and "proof" unless explicitly stated.

In **bookdown**, the types of theorem environments supported are in Table 2.1.
To write a theorem, you can use the syntax below:

```
```{theorem}
Here is my theorem.
```
```

TABLE 2.1: Theorem environments in **bookdown**.

Environment	Printed Name	Label Prefix
theorem	Theorem	thm
lemma	Lemma	lem
definition	Definition	def
corollary	Corollary	cor
proposition	Proposition	prp
example	Example	ex

To write other theorem environments, replace ```` ```{theorem} ```` with other environment names in Table 2.1, e.g., ```` ```{lemma} ````.

A theorem can have a `name` option so its name will be printed, e.g.,

```
```{theorem, name="Pythagorean theorem"}
For a right triangle, if $c$ denotes the length of the hypotenuse
and $a$ and $b$ denote the lengths of the other two sides, we have
$$a^2 + b^2 = c^2$$
```
```

If you want to refer to a theorem, you should label it. The label can be written after ```` ```{theorem ````, e.g.,

```
```{theorem, label="foo"}
A labeled theorem here.
```
```

The `label` option can be implicit, e.g., the following theorem has the label bar:

```
```{theorem, bar}
A labeled theorem here.
```
```

After you label a theorem, you can refer to it using the syntax `\@ref(prefix:label)`. See the column `Label Prefix` in Table 2.1 for the value of `prefix` for each environment. For example, we have a labeled and named theorem below, and `\@ref(thm:pyth)` gives us its theorem number 2.1:

```
```{theorem, pyth, name="Pythagorean theorem"}
For a right triangle, if $c$ denotes the length of the hypotenuse
and $a$ and $b$ denote the lengths of the other two sides, we have

$$a^2 + b^2 = c^2$$
```
```

Theorem 2.1 (Pythagorean theorem). *For a right triangle, if c denotes the length of the hypotenuse and a and b denote the lengths of the other two sides, we have*

$$a^2 + b^2 = c^2$$

The proof environments currently supported are `proof` and `remark`. The syntax is similar to theorem environments, and proof environments can also be named. The only difference is that since they are unnumbered, you cannot reference them.

We have tried to make all these theorem and proof environments work out of the box, no matter if your output is PDF, HTML, or EPUB. If you are a LaTeX or HTML expert, you may want to customize the style of these environments anyway (see Chapter 4). Customization in HTML is easy with CSS, and each environment is enclosed in `<div></div>` with the CSS class being the environment name, e.g., `<div class="lemma"></div>`. For LaTeX output, we have predefined the style to be `definition` for environments `definition` and `example`, and `remark` for environments `proof` and `remark`. All other environments use the `plain` style. The style definition is done through the `\theoremstyle{}` command of the **amsthm** package.

Theorems are numbered by chapters by default. If there are no chapters in your document, they are numbered by sections instead. If the whole document is unnumbered (the output format option `number_sections = FALSE`), all theorems are numbered sequentially from 1, 2, ..., N. LaTeX supports numbering one theorem environment after another, e.g., let theorems and lemmas share the same counter. This is not supported for HTML/EPUB output

in **bookdown**. You can change the numbering scheme in the LaTeX preamble by defining your own theorem environments, e.g.,

```
\newtheorem{theorem}{Theorem}
\newtheorem{lemma}[theorem]{Lemma}
```

When **bookdown** detects \newtheorem{theorem} in your LaTeX preamble, it
will not write out its default theorem definitions, which means you have
to define all theorem environments by yourself. For the sake of simplicity
and consistency, we do not recommend that you do this. It can be confus-
ing when your Theorem 18 in PDF becomes Theorem 2.4 in HTML.

Below we show more examples[2] of the theorem and proof environments, so
you can see the default styles in **bookdown**.

Definition 2.1. The characteristic function of a random variable X is defined by

$$\varphi_X(t) = \mathrm{E}\left[e^{itX}\right], \ t \in \mathcal{R}$$

Example 2.1. We derive the characteristic function of $X \sim U(0,1)$ with
the probability density function $f(x) = \mathbf{1}_{x \in [0,1]}$.

$$\varphi_X(t) = \mathrm{E}\left[e^{itX}\right] = \int e^{itx} f(x) dx = \int_0^1 e^{itx} dx$$

$$= \int_0^1 \left(\cos(tx) + i\sin(tx)\right) dx$$

$$= \left(\frac{\sin(tx)}{t} - i\frac{\cos(tx)}{t}\right)\Big|_0^1$$

$$= \frac{\sin(t)}{t} - i\left(\frac{\cos(t) - 1}{t}\right)$$

$$= \frac{i\sin(t)}{it} + \frac{\cos(t) - 1}{it}$$

$$= \frac{e^{it} - 1}{it}$$

---

[2]Some examples are adapted from the Wikipedia page https://en.wikipedia.org/
wiki/Characteristic_function_(probability_theory)

Note that we used the fact $e^{ix} = \cos(x) + i\sin(x)$ twice.

Lemma 2.1. *For any two random variables X_1, X_2, they both have the same probability distribution if and only if*

$$\varphi_{X_1}(t) = \varphi_{X_2}(t)$$

Theorem 2.2. *If X_1, ..., X_n are independent random variables, and a_1, ..., a_n are some constants, then the characteristic function of the linear combination $S_n = \sum_{i=1}^{n} a_i X_i$ is*

$$\varphi_{S_n}(t) = \prod_{i=1}^{n} \varphi_{X_i}(a_i t) = \varphi_{X_1}(a_1 t) \cdots \varphi_{X_n}(a_n t)$$

Proposition 2.1. The distribution of the sum of independent Poisson random variables $X_i \sim \text{Pois}(\lambda_i)$, $i = 1, 2, \cdots, n$ is $\text{Pois}(\sum_{i=1}^{n} \lambda_i)$.

Proof. The characteristic function of $X \sim \text{Pois}(\lambda)$ is $\varphi_X(t) = e^{\lambda(e^{it}-1)}$. Let $P_n = \sum_{i=1}^{n} X_i$. We know from Theorem 2.2 that

$$\varphi_{P_n}(t) = \prod_{i=1}^{n} \varphi_{X_i}(t)$$

$$= \prod_{i=1}^{n} e^{\lambda_i(e^{it}-1)}$$

$$= e^{\sum_{i=1}^{n} \lambda_i(e^{it}-1)}$$

This is the characteristic function of a Poisson random variable with the parameter $\lambda = \sum_{i=1}^{n} \lambda_i$. From Lemma 2.1, we know the distribution of P_n is $\text{Pois}(\sum_{i=1}^{n} \lambda_i)$. $\square$

Remark. In some cases, it is very convenient and easy to figure out the distribution of the sum of independent random variables using characteristic functions.

Corollary 2.1. The characteristic function of the sum of two independent random variables X_1 and X_2 is the product of characteristic functions of X_1 and X_2, i.e.,

$$\varphi_{X_1+X_2}(t) = \varphi_{X_1}(t)\varphi_{X_2}(t)$$

2.2.3   Special headers

There are a few special types of first-level headers that will be processed dif-
ferently in **bookdown**. The first type is an unnumbered header that starts
with the token (PART). This kind of headers are translated to part titles. If
you are familiar with LaTeX, this basically means \part{}. When your book
has a large number of chapters, you may want to organize them into parts,
e.g.,

```
# (PART) Part I {-}

# Chapter One

# Chapter Two

# (PART) Part II {-}

# Chapter Three
```

A part title should be written right before the first chapter title in this part.

The second type is an unnumbered header that starts with (APPENDIX), indi-
cating that all chapters after this header are appendices, e.g.,

```
# Chapter One

# Chapter Two

# (APPENDIX) Appendix {-}

# Appendix A

# Appendix B
```

The numbering style of appendices will be automatically changed in La-
TeX/PDF and HTML output (usually in the form A, A.1, A.2, B, B.1, ...). This
feature is not available to e-books or Word output.

2.2.4 Text references

You can assign some text to a label and reference the text using the label elsewhere in your document. This can be particularly useful for long figure/table captions (Section 2.4 and 2.5), in which case you normally will have to write the whole character string in the chunk header (e.g., `fig.cap` = `"A long long figure caption."`) or your R code (e.g., `kable(caption` = `"A long long table caption."))`. It is also useful when these captions contain special HTML or LaTeX characters, e.g., if the figure caption contains an underscore, it works in the HTML output but may not work in LaTeX output because the underscore must be escaped in LaTeX.

The syntax for a text reference is `(ref:label)` text, where `label` is a unique label[3] throughout the document for `text`. It must be in a separate paragraph with empty lines above and below it. For example,

```
(ref:foo) Define a text reference **here**.
```

Then you can use `(ref:foo)` in your figure/table captions. The text can contain anything that Markdown supports, as long as it is one single paragraph. Here is a complete example:

```
A normal paragraph.

(ref:foo) A scatterplot of the data `cars` using **base** R graphics.

```{r foo, fig.cap='(ref:foo)'}
plot(cars)  # a scatterplot
```
```

Text references can be used anywhere in the document (not limited to figure captions). It can also be useful if you want to reuse a fragment of text in multiple places.

---

[3]You may consider using the code chunk labels.

2.3   R code

There are two types of R code in R Markdown/**knitr** documents: R code chunks, and inline R code. The syntax for the latter is `` `r R_CODE` ``, and it can be embedded inline with other document elements. R code chunks look like plain code blocks, but have {r} after the three backticks and (optionally) chunk options inside {}, e.g.,

```
```{r chunk-label, echo = FALSE, fig.cap = 'A figure caption.'}
1 + 1
rnorm(10)  # 10 random numbers
plot(dist ~ speed, cars)  # a scatterplot
```
```

To learn more about **knitr** chunk options, see Xie (2015) or the web page http://yihui.name/knitr/options. For books, additional R code can be executed before/after each chapter; see before_chapter_script and after_chapter_script in Section 4.4.

2.4   Figures

By default, figures have no captions in the output generated by **knitr**, which means they will be placed wherever they were generated in the R code. Below is such an example.

```
par(mar = c(4, 4, 0.1, 0.1))
plot(pressure, pch = 19, type = "b")
```

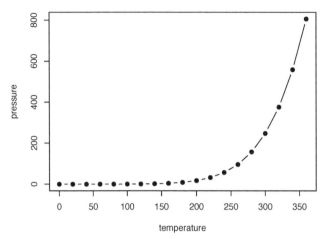

The disadvantage of typesetting figures in this way is that when there is not enough space on the current page to place a figure, it may either reach the bottom of the page (hence exceeds the page margin), or be pushed to the next page, leaving a large white margin at the bottom of the current page. That is basically why there are "floating environments" in LaTeX: elements that cannot be split over multiple pages (like figures) are put in floating environments, so they can float to a page that has enough space to hold them. There is also a disadvantage of floating things forward or backward, though. That is, readers may have to jump to a different page to find the figure mentioned on the current page. This is simply a natural consequence of having to typeset things on multiple pages of fixed sizes. This issue does not exist in HTML, however, since everything can be placed continuously on one single page (presumably with infinite height), and there is no need to split anything across multiple pages of the same page size.

If we assign a figure caption to a code chunk via the chunk option `fig.cap`, R plots will be put into figure environments, which will be automatically labeled and numbered, and can also be cross-referenced. The label of a figure environment is generated from the label of the code chunk, e.g., if the chunk label is `foo`, the figure label will be `fig:foo` (the prefix `fig:` is added before `foo`). To reference a figure, use the syntax `\@ref(label)`,[4] where `label` is the figure label, e.g., `fig:foo`.

---

[4]Do not forget the leading backslash! And also note the parentheses `()` after `ref`; they are not curly braces `{}`.

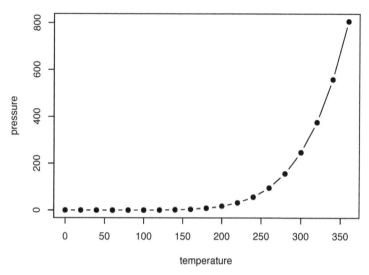

FIGURE 2.1: A figure example with the specified aspect ratio, width, and alignment.

 If you want to cross-reference figures or tables generated from a code chunk, please make sure the chunk label only contains *alphanumeric* characters (a-z, A-Z, 0-9), slashes (/), or dashes (-).

The chunk option `fig.asp` can be used to set the aspect ratio of plots, i.e., the ratio of figure height/width. If the figure width is 6 inches (`fig.width = 6`) and `fig.asp = 0.7`, the figure height will be automatically calculated from `fig.width * fig.asp = 6 * 0.7 = 4.2`. Figure 2.1 is an example using the chunk options `fig.asp = 0.7`, `fig.width = 6`, and `fig.align = 'center'`, generated from the code below:

```
par(mar = c(4, 4, 0.1, 0.1))
plot(pressure, pch = 19, type = "b")
```

The actual size of a plot is determined by the chunk options `fig.width` and `fig.height` (the size of the plot generated from a graphical device), and we can specify the output size of plots via the chunk options `out.width` and `out.height`. The possible value of these two options depends on the output format of the document. For example, `out.width = '30%'` is a valid value

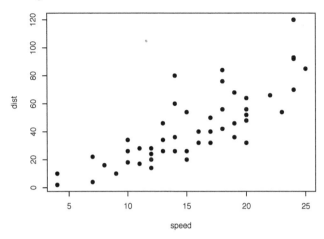

FIGURE 2.2: A figure example with a relative width 70%.

for HTML output, but not for LaTeX/PDF output. However, **knitr** will automatically convert a percentage value for out.width of the form x% to (x / 100) \linewidth, e.g., out.width = '70%' will be treated as .7\linewidth when the output format is LaTeX. This makes it possible to specify a relative width of a plot in a consistent manner. Figure 2.2 is an example of out.width = 70%.

```
par(mar = c(4, 4, 0.1, 0.1))
plot(cars, pch = 19)
```

If you want to put multiple plots in one figure environment, you must use the chunk option fig.show = 'hold' to hold multiple plots from a code chunk and include them in one environment. You can also place plots side by side if the sum of the width of all plots is smaller than or equal to the current line width. For example, if two plots have the same width 50%, they will be placed side by side. Similarly, you can specify out.width = '33%' to arrange three plots on one line. Figure 2.3 is an example of two plots, each with a width of 50%.

```
par(mar = c(4, 4, 0.1, 0.1))
plot(pressure, pch = 19, type = "b")
plot(cars, pch = 19)
```

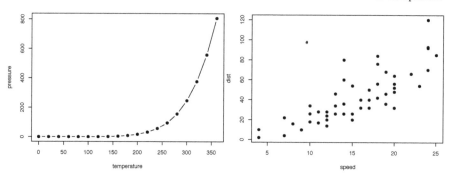

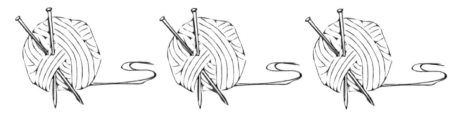

FIGURE 2.3: Two plots placed side by side.

FIGURE 2.4: Three knitr logos included in the document from an external PNG image file.

Sometimes you may have certain images that are not generated from R code, and you can include them in R Markdown via the function `knitr::include_graphics()`. Figure 2.4 is an example of three **knitr** logos included in a figure environment. You may pass one or multiple image paths to the `include_graphics()` function, and all chunk options that apply to normal R plots also apply to these images, e.g., you can use `out.width = '33%'` to set the widths of these images in the output document.

```
knitr::include_graphics(rep("images/knit-logo.png", 3))
```

There are a few advantages of using `include_graphics()`:

1. You do not need to worry about the document output format, e.g., when the output format is LaTeX, you may have to use the LaTeX command `\includegraphics{}` to include an image, and when the output format is Markdown, you have to use `![]()`. The function

include_graphics() in **knitr** takes care of these details automatically.

2. The syntax for controlling the image attributes is the same as when images are generated from R code, e.g., chunk options fig.cap, out.width, and fig.show still have the same meanings.

3. include_graphics() is smart enough to use PDF graphics automatically when the output format is LaTeX and the PDF graphics files exist, e.g., an image path foo/bar.png can be automatically replaced with foo/bar.pdf if the latter exists. PDF images often have better qualities than raster images in LaTeX/PDF output. Of course, you can disable this feature by include_graphics(auto_pdf = FALSE) if you do not like it.

4. You can easily scale these images proportionally using the same ratio. This can be done via the dpi argument (dots per inch), which takes the value from the chunk option dpi by default. If it is a numeric value and the chunk option out.width is not set, the output width of an image will be its actual width (in pixels) divided by dpi, and the unit will be inches. For example, for an image with the size 672 x 480, its output width will be 7 inches (7in) when dpi = 96. This feature requires the package **png** and/or **jpeg** to be installed. You can always override the automatic calculation of width in inches by providing a non-NULL value to the chunk option out.width, or use include_graphics(dpi = NA).

2.5 Tables

For now, the most convenient way to generate a table is the function knitr::kable(), because there are some internal tricks in **knitr** to make it work with **bookdown** and users do not have to know anything about these implementation details. We will explain how to use other packages and functions later in this section.

Like figures, tables with captions will also be numbered and can be referenced. The kable() function will automatically generate a label for a table environment, which is the prefix tab: plus the chunk label. For example, the

TABLE 2.2: A table of the first 10 rows of the mtcars data.

	mpg	cyl	disp	hp	drat	wt	qsec	vs
Mazda RX4	21.0	6	160.0	110	3.90	2.620	16.46	0
Mazda RX4 Wag	21.0	6	160.0	110	3.90	2.875	17.02	0
Datsun 710	22.8	4	108.0	93	3.85	2.320	18.61	1
Hornet 4 Drive	21.4	6	258.0	110	3.08	3.215	19.44	1
Hornet Sportabout	18.7	8	360.0	175	3.15	3.440	17.02	0
Valiant	18.1	6	225.0	105	2.76	3.460	20.22	1
Duster 360	14.3	8	360.0	245	3.21	3.570	15.84	0
Merc 240D	24.4	4	146.7	62	3.69	3.190	20.00	1
Merc 230	22.8	4	140.8	95	3.92	3.150	22.90	1
Merc 280	19.2	6	167.6	123	3.92	3.440	18.30	1

table label for a code chunk with the label foo will be tab:foo, and we can still use the syntax \@ref(label) to reference the table. Table 2.2 is a simple example.

```
knitr::kable(
  head(mtcars[, 1:8], 10), booktabs = TRUE,
  caption = 'A table of the first 10 rows of the mtcars data.'
)
```

If you want to put multiple tables in a single table environment, wrap the data objects (usually data frames in R) into a list. See Table 2.3 for an example.

```
knitr::kable(
  list(
    head(iris[, 1:2], 3),
    head(mtcars[, 1:3], 5)
  ),
  caption = 'A Tale of Two Tables.', booktabs = TRUE
)
```

When you do not want a table to float in PDF, you may use the LaTeX

TABLE 2.3: A Tale of Two Tables.

Sepal.Length	Sepal.Width		mpg	cyl	disp
5.1	3.5	Mazda RX4	21.0	6	160
4.9	3.0	Mazda RX4 Wag	21.0	6	160
4.7	3.2	Datsun 710	22.8	4	108
		Hornet 4 Drive	21.4	6	258
		Hornet Sportabout	18.7	8	360

package **longtable**,[5] which can break a table across multiple pages. To use **longtable**, pass longtable = TRUE to kable(), and make sure to include \usepackage{longtable} in the LaTeX preamble (see Section 4.1 for how to customize the LaTeX preamble). Of course, this is irrelevant to HTML output, since tables in HTML do not need to float.

```
knitr::kable(
  iris[1:60, ], longtable = TRUE, booktabs = TRUE,
  caption = 'A table generated by the longtable package.'
)
```

TABLE 2.4: A table generated by the longtable package.

Sepal.Length	Sepal.Width	Petal.Length	Petal.Width	Species
5.1	3.5	1.4	0.2	setosa
4.9	3.0	1.4	0.2	setosa
4.7	3.2	1.3	0.2	setosa
4.6	3.1	1.5	0.2	setosa
5.0	3.6	1.4	0.2	setosa
5.4	3.9	1.7	0.4	setosa
4.6	3.4	1.4	0.3	setosa
5.0	3.4	1.5	0.2	setosa
4.4	2.9	1.4	0.2	setosa
4.9	3.1	1.5	0.1	setosa
5.4	3.7	1.5	0.2	setosa

[5]https://www.ctan.org/pkg/longtable

4.8	3.4	1.6	0.2	setosa
4.8	3.0	1.4	0.1	setosa
4.3	3.0	1.1	0.1	setosa
5.8	4.0	1.2	0.2	setosa
5.7	4.4	1.5	0.4	setosa
5.4	3.9	1.3	0.4	setosa
5.1	3.5	1.4	0.3	setosa
5.7	3.8	1.7	0.3	setosa
5.1	3.8	1.5	0.3	setosa
5.4	3.4	1.7	0.2	setosa
5.1	3.7	1.5	0.4	setosa
4.6	3.6	1.0	0.2	setosa
5.1	3.3	1.7	0.5	setosa
4.8	3.4	1.9	0.2	setosa
5.0	3.0	1.6	0.2	setosa
5.0	3.4	1.6	0.4	setosa
5.2	3.5	1.5	0.2	setosa
5.2	3.4	1.4	0.2	setosa
4.7	3.2	1.6	0.2	setosa
4.8	3.1	1.6	0.2	setosa
5.4	3.4	1.5	0.4	setosa
5.2	4.1	1.5	0.1	setosa
5.5	4.2	1.4	0.2	setosa
4.9	3.1	1.5	0.2	setosa
5.0	3.2	1.2	0.2	setosa
5.5	3.5	1.3	0.2	setosa
4.9	3.6	1.4	0.1	setosa
4.4	3.0	1.3	0.2	setosa
5.1	3.4	1.5	0.2	setosa
5.0	3.5	1.3	0.3	setosa
4.5	2.3	1.3	0.3	setosa
4.4	3.2	1.3	0.2	setosa
5.0	3.5	1.6	0.6	setosa
5.1	3.8	1.9	0.4	setosa

4.8	3.0	1.4	0.3	setosa
5.1	3.8	1.6	0.2	setosa
4.6	3.2	1.4	0.2	setosa
5.3	3.7	1.5	0.2	setosa
5.0	3.3	1.4	0.2	setosa
7.0	3.2	4.7	1.4	versicolor
6.4	3.2	4.5	1.5	versicolor
6.9	3.1	4.9	1.5	versicolor
5.5	2.3	4.0	1.3	versicolor
6.5	2.8	4.6	1.5	versicolor
5.7	2.8	4.5	1.3	versicolor
6.3	3.3	4.7	1.6	versicolor
4.9	2.4	3.3	1.0	versicolor
6.6	2.9	4.6	1.3	versicolor
5.2	2.7	3.9	1.4	versicolor

If you decide to use other packages to generate tables, you have to make sure the label for the table environment appears in the beginning of the table caption in the form (\#label), where label must have the prefix tab:. You have to be very careful about the *portability* of the table generating function: it should work for both HTML and LaTeX output automatically, so it must consider the output format internally (check knitr::opts_knit$get('pandoc.to')). When writing out an HTML table, the caption must be written in the <caption></caption> tag. For simple tables, kable() should suffice. If you have to create complicated tables (e.g., with certain cells spanning across multiple columns/rows), you will have to take the aforementioned issues into consideration.

2.6 Cross-references

We have explained how cross-references work for equations (Section 2.2.1), theorems (Section 2.2.2), figures (Section 2.4), and tables (Section 2.5). In fact, you can also reference sections using the same syntax \@ref(label), where label is the section ID. By default, Pandoc will generate an ID for all section headers, e.g., a section `# Hello World` will have an ID `hello-world`. We recommend you to manually assign an ID to a section header to make sure you do not forget to update the reference label after you change the section header. To assign an ID to a section header, simply add `{#id}` to the end of the section header.

When a referenced label cannot be found, you will see two question marks like **??**, as well as a warning message in the R console when rendering the book.

You can also create text-based links using explicit or automatic section IDs or even the actual section header text.

- If you are happy with the section header as the link text, use it inside a single set of square brackets:
    - `[Section header text]`: example "A single document" via `[A single document]`
- There are two ways to specify custom link text:
    - `[link text][Section header text]`, e.g., "non-English books" via `[non-English books][Internationalization]`
    - `[link text](#ID)`, e.g., "Table stuff" via `[Table stuff](#tables)`

The Pandoc documentation provides more details on automatic section IDs[6] and implicit header references.[7]

Cross-references still work even when we refer to an item that is not on the current page of the PDF or HTML output. For example, see Equation (2.1) and Figure 2.4.

2.7 Custom blocks

You can generate custom blocks using the `block` engine in **knitr**, i.e., the chunk option `engine = 'block'`, or the more compact syntax ```` ```{block} ````.

---

[6]http://pandoc.org/MANUAL.html#extension-auto_identifiers
[7]http://pandoc.org/MANUAL.html#extension-implicit_header_references

This engine should be used in conjunction with the chunk option `type`, which takes a character string. When the `block` engine is used, it generates a `<div>` to wrap the chunk content if the output format is HTML, and a LaTeX environment if the output is LaTeX. The `type` option specifies the class of the `<div>` and the name of the LaTeX environment. For example, the HTML output of this chunk

```
```{block, type='FOO'}
Some text for this block.
```
```

will be this:

```
<div class="FOO">
Some text for this block.
</div>
```

and the LaTeX output will be this:

```
\begin{FOO}
Some text for this block.
\end{FOO}
```

It is up to the book author how to define the style of the block. You can define the style of the `<div>` in CSS and include it in the output via the `includes` option in the YAML metadata. Similarly, you may define the LaTeX environment via `\newenvironment` and include the definition in the LaTeX output via the `includes` option. For example, we may save the following style in a CSS file, say, `style.css`:

```
div.FOO {
  font-weight: bold;
  color: red;
}
```

And the YAML metadata of the R Markdown document can be:

```
---
output:
  bookdown::html_book:
    includes:
      in_header: style.css
---
```

We have defined a few types of blocks for this book to show notes, tips, and warnings, etc. Below are some examples:

R is free software and comes with ABSOLUTELY NO WARRANTY. You are welcome to redistribute it under the terms of the GNU General Public License versions 2 or 3. For more information about these matters see http://www.gnu.org/licenses/.

R is free software and comes with ABSOLUTELY NO WARRANTY. You are welcome to redistribute it under the terms of the GNU General Public License versions 2 or 3. For more information about these matters see http://www.gnu.org/licenses/.

R is free software and comes with ABSOLUTELY NO WARRANTY. You are welcome to redistribute it under the terms of the GNU General Public License versions 2 or 3. For more information about these matters see http://www.gnu.org/licenses/.

R is free software and comes with ABSOLUTELY NO WARRANTY. You are welcome to redistribute it under the terms of the GNU General Public License versions 2 or 3. For more information about these matters see http://www.gnu.org/licenses/.

R is free software and comes with ABSOLUTELY NO WARRANTY. You are welcome to redistribute it under the terms of the GNU General Pub-

lic License versions 2 or 3. For more information about these matters see
`http://www.gnu.org/licenses/`.

The **knitr** `block` engine was designed to display simple content (typically a
paragraph of plain text). You can use simple formatting syntax such as mak-
ing certain words bold or italic, but more advanced syntax such as citations
and cross-references will not work. However, there is an alternative engine
named `block2` that supports arbitrary Markdown syntax, e.g.,

```
```{block2, type='FOO'}
Some text for this block [@citation-key].

- a list item
- another item

More text.
```
```

The `block2` engine should also be faster than the `block` engine if you have a
lot of custom blocks in the document, but its implementation was based on
a hack,[8] so we are not 100% sure if it is always going to work in the future.
We have not seen problems with Pandoc v1.17.2 yet.

One more caveat for the `block2` engine: if the last element in the block is not
an ordinary paragraph, you must leave a blank line at the end, e.g.,

```
```{block2, type='FOO'}
Some text for this block [@citation-key].

- a list item
- another item
- end the list with a blank line

```
```

---

[8]`https://github.com/jgm/pandoc/issues/2453`

The theorem and proof environments in Section 2.2.2 are actually implemented through the `block2` engine.

For all custom blocks based on the `block` or `block2` engine, there is one chunk option `echo` that you can use to show (`echo = TRUE`) or hide (`echo = FALSE`) the blocks.

2.8 Citations

Although Pandoc supports multiple ways of writing citations, we recommend you to use BibTeX databases because they work best with LaTeX/PDF output. Pandoc can process other types of bibliography databases with the utility `pandoc-citeproc` (`https://github.com/jgm/pandoc-citeproc`), but it may not render certain bibliography items correctly (especially in case of multiple authors per item), and BibTeX can do a better job when the output format is LaTeX. With BibTeX databases, you will be able to define the bibliography style if it is required by a certain publisher or journal.

A BibTeX database is a plain-text file (with the conventional filename extension `.bib`) that consists of bibliography entries like this:

```
@Manual{R-base,
    title = {R: A Language and Environment for Statistical
      Computing},
    author = {{R Core Team}},
    organization = {R Foundation for Statistical Computing},
    address = {Vienna, Austria},
    year = {2016},
    url = {https://www.R-project.org/},
}
```

A bibliography entry starts with `@type{`, where `type` may be `article`, `book`, `manual`, and so on.[9] Then there is a citation key, like `R-base` in the above exam-

---

[9]The type name is case-insensitive, so it does not matter if it is `manual`, `Manual`, or `MANUAL`.

ple. To cite an entry, use @key or [@key] (the latter puts the citation in braces),
e.g., @R-base is rendered as R Core Team (2016), and [@R-base] generates "(R
Core Team, 2016)". If you are familiar with the **natbib** package in LaTeX, @key
is basically \citet{key}, and [@key] is equivalent to \citep{key}.

There are a number of fields in a bibliography entry, such as title, author,
and year, etc. You may see https://en.wikipedia.org/wiki/BibTeX for possible types of entries and fields in BibTeX.

There is a helper function write_bib() in **knitr** to generate BibTeX entries
automatically for R packages. Note that it only generates one BibTeX entry
for the package itself at the moment, whereas a package may contain multiple entries in the CITATION file, and some entries are about the publications
related to the package. These entries are ignored by write_bib().

```
# the second argument can be a .bib file
knitr::write_bib(c("knitr", "stringr"), "", width = 60)
```

```
@Manual{R-knitr,
    title = {knitr: A General-Purpose Package for Dynamic Report
        Generation in R},
    author = {Yihui Xie},
    year = {2016},
    note = {R package version 1.15},
    url = {http://yihui.name/knitr/},
}
@Manual{R-stringr,
    title = {stringr: Simple, Consistent Wrappers for Common
        String Operations},
    author = {Hadley Wickham},
    year = {2016},
    note = {R package version 1.1.0},
    url = {https://CRAN.R-project.org/package=stringr},
}
```

Once you have one or multiple .bib files, you may use the field bibliography
in the YAML metadata of your R Markdown document, and you can also

specify the bibliography style via `biblio-style` (this only applies to PDF output), e.g.,

```
---
bibliography: ["one.bib", "another.bib", "yet-another.bib"]
biblio-style: "apalike"
link-citations: true
---
```

The field `link-citations` can be used to add internal links from the citation text of the author-year style to the bibliography entry in the HTML output.

When the output format is LaTeX, citations will be automatically put in a chapter or section. For non-LaTeX output, you can add an empty chapter as the last chapter of your book. For example, if your last chapter is the Rmd file `06-references.Rmd`, its content can be an inline R expression:

```
`r if (knitr:::is_html_output()) '# References {-}'`
```

2.9   Index

Currently the index is only supported for LaTeX/PDF output. To print an index after the book, you can use the LaTeX package **makeidx** in the preamble (see Section 4.1):

```
\usepackage{makeidx}
\makeindex
```

Then insert `\printindex` at the end of your book through the YAML option `includes -> after_body`. An index entry can be created via the `\index{}` command in the book body, e.g., `\index{GIT}`.

2.10   HTML widgets

Although one of R's greatest strengths is data visualization, there are a large number of JavaScript libraries for much richer data visualization. These libraries can be used to build interactive applications that can easily render in web browsers, so users do not need to install any additional software packages to view the visualizations. One way to bring these JavaScript libraries into R is through the **htmlwidgets**[10] package (Vaidyanathan et al., 2016).

HTML widgets can be rendered as a standalone web page (like an R plot), or embedded in R Markdown documents and Shiny applications. They were originally designed for HTML output only, and they require the availability of JavaScript, so they will not work in non-HTML output formats, such as LaTeX/PDF. Before **knitr** v1.13, you will get an error when you render HTML widgets to an output format that is not HTML. Since **knitr** v1.13, HTML widgets will be rendered automatically as screenshots taken via the **webshot** package (Chang, 2016). Of course, you need to install the **webshot** package. Additionally, you have to install PhantomJS (`http://phantomjs.org`), since it is what **webshot** uses to capture screenshots. Both **webshot** and PhantomJS can be installed automatically from R:

```
install.packages("webshot")
webshot::install_phantomjs()
```

The function `install_phantomjs()` works for Windows, OS X, and Linux. You may also choose to download and install PhantomJS by yourself, if you are familiar with modifying the system environment variable PATH.

When **knitr** detects an HTML widget object in a code chunk, it either renders the widget normally when the current output format is HTML, or saves the widget as an HTML page and calls **webshot** to capture the screen of the HTML page when the output format is not HTML. Here is an example of a table created from the **DT** package (Xie, 2016b):

---

[10]`http://htmlwidgets.org`

```
DT::datatable(iris)
```

Show 10 entries				Search:
	Sepal.Length	Sepal.Width	Petal.Length	Petal.Width	Species
1	5.1	3.5	1.4	0.2	setosa
2	4.9	3	1.4	0.2	setosa
3	4.7	3.2	1.3	0.2	setosa
4	4.6	3.1	1.5	0.2	setosa
5	5	3.6	1.4	0.2	setosa
6	5.4	3.9	1.7	0.4	setosa
7	4.6	3.4	1.4	0.3	setosa
8	5	3.4	1.5	0.2	setosa
9	4.4	2.9	1.4	0.2	setosa
10	4.9	3.1	1.5	0.1	setosa

Showing 1 to 10 of 150 entries   Previous   1   2   3   4   5   ...   15   Next

FIGURE 2.5: A table widget rendered via the DT package.

If you are reading this book as web pages now, you should see an interactive table generated from the above code chunk, e.g., you may sort the columns and search in the table. If you are reading a non-HTML version of this book, you should see a screenshot of the table. The screenshot may look a little different with the actual widget rendered in the web browser, due to the difference between a real web browser and PhantomJS's virtual browser.

There are a number of **knitr** chunk options related to screen-capturing. First, if you are not satisfied with the quality of the automatic screenshots, or want a screenshot of the widget of a particular state (e.g., after you click and sort a certain column of a table), you may capture the screen manually, and provide your own screenshot via the chunk option screenshot.alt (alternative screenshots). This option takes the paths of images. If you have multiple widgets in a chunk, you can provide a vector of image paths. When this option is present, **knitr** will no longer call **webshot** to take automatic screenshots.

Second, sometimes you may want to force **knitr** to use static screenshots instead of rendering the actual widgets even on HTML pages. In this case, you can set the chunk option screenshot.force = TRUE, and widgets will always be rendered as static images. Note that you can still choose to use automatic or custom screenshots.

Third, **webshot** has some options to control the automatic screenshots, and you may specify these options via the chunk option screenshot.opts, which takes a list like list(delay = 2, cliprect = 'viewport'). See the help page ?webshot::webshot for the full list of possible options, and the package vignette[11] vignette('intro', package = 'webshot') illustrates the effect of these options. Here the delay option can be important for widgets that take long time to render: delay specifies the number of seconds to wait before PhantomJS takes the screenshot. If you see an incomplete screenshot, you may want to specify a longer delay (the default is 0.2 seconds).

Fourth, if you feel it is slow to capture the screenshots, or do not want to do it every time the code chunk is executed, you may use the chunk option cache = TRUE to cache the chunk. Caching works for both HTML and non-HTML output formats.

Screenshots behave like normal R plots in the sense that many chunk options related to figures also apply to screenshots, including fig.width, fig.height, out.width, fig.cap, and so on. So you can specify the size of screenshots in the output document, and assign figure captions to them as well. The image format of the automatic screenshots can be specified via the chunk option dev, and possible values are pdf, png, and jpeg. The default for PDF output is pdf, and it is png for other types of output. Note that pdf may not work as faithfully as png: sometimes there are certain elements on an HTML page that fail to render to the PDF screenshot, so you may want to use dev = 'png' even for PDF output. It depends on specific cases of HTML widgets, and you can try both pdf and png (or jpeg) before deciding which format is more desirable.

2.11 Web pages and Shiny apps

Similar to HTML widgets, arbitrary web pages can be embedded in the book. You can use the function knitr::include_url() to include a web page through its URL. When the output format is HTML, an iframe is used;[12] in

---

[11]https://cran.rstudio.com/web/packages/webshot/vignettes/intro.html

[12]An iframe is basically a box on one web page to embed another web page.

other cases, **knitr** tries to take a screenshot of the web page (or use the custom screenshot you provided). All chunk options are the same as those for HTML widgets. One option that may require your special attention is the delay option: HTML widgets are rendered locally, so usually they are fast to load for PhantomJS to take screenshots, but an arbitrary URL may take longer to load, so you may want to use a larger delay value, e.g., use the chunk option screenshot.opts = list(delay = 5).

A related function is knitr::include_app(), which is very similar to include_url(), and it was designed for embedding Shiny apps via their URLs in the output. Its only difference with include_url() is that it automatically adds a query parameter ?showcase=0 to the URL, if no other query parameters are present in the URL, to disable the Shiny showcase mode, which is unlikely to be useful for screenshots or iframes. If you do want the showcase mode, use include_url() instead of include_app(). Below is a Shiny app example (Figure 2.6):

```
knitr::include_app("https://yihui.shinyapps.io/miniUI/",
  height = "600px")
```

Again, you will see a live app if you are reading an HTML version of this book, and a static screenshot if you are reading other types of formats. The above Shiny app was created using the **miniUI** package (Cheng, 2016), which provides layout functions that are particularly nice for Shiny apps on small screens. If you use normal Shiny layout functions, you are likely to see vertical and/or horizontal scrollbars in the iframes because the page size is too big to fit an iframe. When the default width of the iframe is too small, you may use the chunk option out.width to change it. For the height of the iframe, use the height argument of include_url()/include_app().

Shiny apps may take even longer to load than usual URLs. You may want to use a conservative value for the delay option, e.g., 10. Needless to say, include_url() and include_app() require a working Internet connection, unless you have previously cached the chunk (but web pages inside iframes still will not work without an Internet connection).

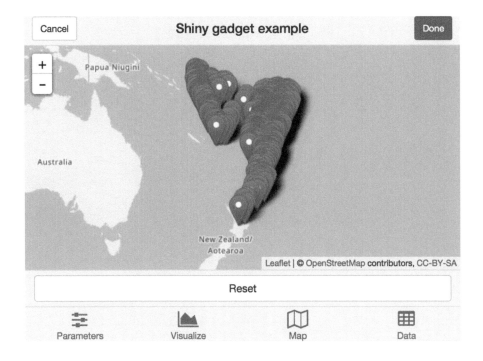

FIGURE 2.6: A Shiny app created via the miniUI package; you can see a live version at https://yihui.shinyapps.io/miniUI/.

3

Output Formats

The **bookdown** package primarily supports three types of output formats: HTML, LaTeX/PDF, and e-books. In this chapter, we introduce the possible options for these formats. Output formats can be specified either in the YAML metadata of the first Rmd file of the book, or in a separate YAML file named _output.yml under the root directory of the book. Here is a brief example of the former (output formats are specified in the output field of the YAML metadata):

```
---
title: "An Impressive Book"
author: "Li Lei and Han Meimei"
output:
  bookdown::gitbook:
    lib_dir: assets
    split_by: section
    config:
      toolbar:
        position: static
  bookdown::pdf_book:
    keep_tex: yes
  bookdown::html_book:
    css: toc.css
documentclass: book
---
```

Here is an example of _output.yml:

```
bookdown::gitbook:
  lib_dir: assets
```

```
  split_by: section
  config:
    toolbar:
      position: static
bookdown::pdf_book:
  keep_tex: yes
bookdown::html_book:
  css: toc.css
```

In this case, all formats should be at the top level, instead of under an output field. You do not need the three dashes --- in _output.yml.

3.1   HTML

The main difference between rendering a book (using **bookdown**) with rendering a single R Markdown document (using **rmarkdown**) to HTML is that a book will generate multiple HTML pages by default — normally one HTML file per chapter. This makes it easier to bookmark a certain chapter or share its URL with others as you read the book, and faster to load a book into the web browser. Currently we have provided a number of different styles for HTML output: the GitBook style, the Bootstrap style, and the Tufte style.

3.1.1   GitBook style

The GitBook style was borrowed from GitBook, a project launched by Friendcode, Inc. (https://www.gitbook.com) and dedicated to helping authors write books with Markdown. It provides a beautiful style, with a layout consisting of a sidebar showing the table of contents on the left, and the main body of a book on the right. The design is responsive to the window size, e.g., the navigation buttons are displayed on the left/right of the book body when the window is wide enough, and collapsed into the bottom when the window is narrow to give readers more horizontal space to read the book body.

We have made several improvements over the original GitBook project. The most significant one is that we replaced the Markdown engine with R Markdown v2 based on Pandoc, so that there are a lot more features for you to use when writing a book:

- You can embed R code chunks and inline R expressions in Markdown, and this makes it easy to create reproducible documents and frees you from synchronizing your computation with its actual output (**knitr** will take care of it automatically).
- The Markdown syntax is much richer: you can write anything that Pandoc's Markdown supports, such as LaTeX math expressions and citations.
- You can embed interactive content in the book (for HTML output only), such as HTML widgets and Shiny apps.

We have also added some useful features in the user interface that we will introduce in detail soon. The output format function for the GitBook style in **bookdown** is `gitbook()`. Here are its arguments:

```
gitbook(fig_caption = TRUE, number_sections = TRUE,
  self_contained = FALSE, lib_dir = "libs", ...,
  split_by = c("chapter", "chapter+number", "section",
    "section+number", "rmd", "none"), split_bib = TRUE,
  config = list())
```

Most arguments are passed to `rmarkdown::html_document()`, including `fig_caption`, `lib_dir`, and `....`. You can check out the help page of `rmarkdown::html_document()` for the full list of possible options. We strongly recommend you to use `fig_caption = TRUE` for two reasons: 1) it is important to explain your figures with captions; 2) enabling figure captions means figures will be placed in floating environments when the output is LaTeX, otherwise you may end up with a lot of white space on certain pages. The format of figure/table numbers depends on if sections are numbered or not: if `number_sections = TRUE`, these numbers will be of the format $x.i$, where x is the chapter number, and i in an incremental number; if sections are not numbered, all figures/tables will be numbered sequentially through the book from 1, 2, ..., N. Note that in either case, figures and tables will be numbered independently.

Among all possible arguments in . . ., you are most likely to use the css argument to provide one or more custom CSS files to tweak the default CSS style. There are a few arguments of html_document() that have been hard-coded in gitbook() and you cannot change them: toc = TRUE (there must be a table of contents), theme = NULL (not using any Bootstrap themes), and template (there exists an internal GitBook template).

Please note that if you change self_contained = TRUE to make self-contained HTML pages, the total size of all HTML files can be significantly increased since there are many JS and CSS files that have to be embedded in every single HTML file.

Besides these html_document() options, gitbook() has three other arguments: split_by, split_bib, and config. The split_by argument specifies how you want to split the HTML output into multiple pages, and its possible values are:

- rmd: use the base filenames of the input Rmd files to create the HTML filenames, e.g., generate chapter3.html for chapter3.Rmd.
- none: do not split the HTML file (the book will be a single HTML file).
- chapter: split the file by the first-level headers.
- section: split the file by the second-level headers.
- chapter+number and section+number: similar to chapter and section, but the files will be numbered.

For chapter and section, the HTML filenames will be determined by the header identifiers, e.g., the filename for the first chapter with a chapter title # Introduction will be introduction.html by default. For chapter+number and section+number, the chapter/section numbers will be prepended to the HTML filenames, e.g., 1-introduction.html and 2-1-literature.html. The header identifier is automatically generated from the header text by default,[1] and you can manually specify an identifier using the syntax {#your-custom-id} after the header text, e.g.,

---

[1]To see more details on how an identifier is automatically generated, see the auto_identifiers extension in Pandoc's documentation http://pandoc.org/MANUAL.html#header-identifiers

```
# An Introduction {#introduction}

The default identifier is `an-introduction` but we changed
it to `introduction`.
```

By default, the bibliography is split and relevant citation items are put at the bottom of each page, so that readers do not have to navigate to a different bibliography page to see the details of citations. This feature can be disabled using `split_bib` = FALSE, in which case all citations are put on a separate page.

There are several sub-options in the `config` option for you to tweak some details in the user interface. Recall that all output format options (not only for `bookdown::gitbook`) can be either passed to the format function if you use the command-line interface `bookdown::render_book()`, or written in the YAML metadata. We display the default sub-options of `config` in the `gitbook` format as YAML metadata below (note that they are indented under the `config` option):

```
bookdown::gitbook:
  config:
    toc:
      collapse: subsection
      scroll_highlight: yes
      before: null
      after: null
    toolbar:
      position: fixed
    edit : null
    download: null
    search: yes
    fontsettings:
      theme: white
      family: sans
      size: 2
    sharing:
```

```
facebook: yes
twitter: yes
google: no
weibo: no
instapper: no
vk: no
all: ['facebook', 'google', 'twitter', 'weibo', 'instapaper']
```

The `toc` option controls the behavior of the table of contents (TOC). You can collapse some items initially when a page is loaded via the `collapse` option. Its possible values are `subsection`, `section`, `none` (or `null`). This option can be helpful if your TOC is very long and has more than three levels of headings: `subsection` means collapsing all TOC items for subsections (X.X.X), `section` means those items for sections (X.X) so only the top-level headings are displayed initially, and `none` means not collapsing any items in the TOC. For those collapsed TOC items, you can toggle their visibility by clicking their parent TOC items. For example, you can click a chapter title in the TOC to show/hide its sections.

The `scroll_highlight` option in `toc` indicates whether to enable highlighting of TOC items as you scroll the book body (by default this feature is enabled). Whenever a new header comes into the current viewport as you scroll down/up, the corresponding item in TOC on the left will be highlighted.

Since the sidebar has a fixed width, when an item in the TOC is truncated because the heading text is too wide, you can hover the cursor over it to see a tooltip showing the full text.

You may add more items before and after the TOC using the HTML tag `<li>`. These items will be separated from the TOC using a horizontal divider. You can use the pipe character | so that you do not need to escape any characters in these items following the YAML syntax, e.g.,

```
toc:
  before: |
    <li><a href="...">My Awesome Book</a></li>
    <li><a href="...">John Smith</a></li>
  after: |
```

FIGURE 3.1: The GitBook toolbar.

```
<li><a href="https://github.com/rstudio/bookdown">
Proudly published with bookdown</a></li>
```

As you navigate through different HTML pages, we will try to preserve the scroll position of the TOC. Normally you will see the scrollbar in the TOC at a fixed position even if you navigate to the next page. However, if the TOC item for the current chapter/section is not visible when the page is loaded, we will automatically scroll the TOC to make it visible to you.

The GitBook style has a toolbar (Figure 3.1) at the top of each page that allows you to dynamically change the book settings. The `toolbar` option has a sub-option `position`, which can take values `fixed` or `static`. The default is that the toolbar will be fixed at the top of the page, so even if you scroll down the page, the toolbar is still visible there. If it is `static`, the toolbar will not scroll with the page, i.e., once you scroll away, you will no longer see it.

The first button on the toolbar can toggle the visibility of the sidebar. You can also hit the S key on your keyboard to do the same thing. The GitBook style can remember the visibility status of the sidebar, e.g., if you closed the sidebar, it will remain closed the next time you open the book. In fact, the GitBook style remembers many other settings as well, such as the search keyword and the font settings.

The second button on the toolbar is the search button. Its keyboard shortcut is F (Find). When the button is clicked, you will see a search box at the top of the sidebar. As you type in the box, the TOC will be filtered to display the sections that match the search keyword. Now you can use the arrow keys Up/Down to highlight the next keyword on the current page. When you click the search button again (or hit F outside the search box), the search keyword will be emptied and the search box will be hidden. To disable searching, set the option search: no in config.

The third button is for font/theme settings. You can change the font size (bigger or smaller), the font family (serif or sans serif), and the theme (White, Sepia, or Night). These settings can be changed via the fontsettings option.

The edit option is the same as the option mentioned in Section 4.4. If it is not empty, an edit button will be added to the toolbar. This was designed for potential contributors to the book to contribute by editing the book on GitHub after clicking the button and sending pull requests.

If your book has other output formats for readers to download, you may provide the download option so that a download button can be added to the toolbar. This option takes either a character vector, or a list of character vectors with the length of each vector being 2. When it is a character vector, it should be either a vector of filenames, or filename extensions, e.g., both of the following settings are okay:

```
download: ["book.pdf", "book.epub"]
download: ["pdf", "epub", "mobi"]
```

When you only provide the filename extensions, the filename is derived from the book filename of the configuration file _bookdown.yml (Section 4.4). When download is null, gitbook() will look for PDF, EPUB, and MOBI files in the book output directory, and automatically add them to the download option. If you just want to suppress the download button, use download: no. All files for readers to download will be displayed in a drop-down menu, and the filename extensions are used as the menu text. When the only available format for readers to download is PDF, the download button will be a single PDF button instead of a drop-down menu.

An alternative form for the value of the download option is a list of length-2 vectors, e.g.,

```
download: [["book.pdf", "PDF"], ["book.epub", "EPUB"]]
```

You can also write it as:

```
download:
  - ["book.pdf", "PDF"]
  - ["book.epub", "EPUB"]
```

Each vector in the list consists of the filename and the text to be displayed in the menu. Compared to the first form, this form allows you to customize the menu text, e.g., you may have two different copies of the PDF for readers to download and you will need to make the menu items different.

On the right of the toolbar, there are some buttons to share the link on social network websites such as Twitter, Facebook, and Google+. You can use the `sharing` option to decide which buttons to enable. If you want to get rid of these buttons entirely, use `sharing: null` (or `no`).

Finally, there are a few more top-level options in the YAML metadata that can be passed to the GitBook HTML template via Pandoc. They may not have clear visible effects on the HTML output, but they may be useful when you deploy the HTML output as a website. These options include:

- `description`: A character string to be written to the `content` attribute of the tag `<meta name="description" content="">` in the HTML head (if missing, the title of the book will be used). This can be useful for search engine optimization (SEO). Note that it should be plain text without any Markdown formatting such as `_italic_` or `**bold**`.
- `url`: The URL of book's website, e.g., `https\://bookdown.org/yihui/bookdown/`.[2]
- `github-repo`: The GitHub repository of the book of the form `user/repo`.
- `cover-image`: The path to the cover image of the book.
- `apple-touch-icon`: A path to an icon (e.g., a PNG image). This is for iOS only: when the website is added to the Home screen, the link is represented by this icon.
- `apple-touch-icon-size`: The size of the icon (by default, 152 x 152 pixels).
- `favicon`: A path to the "favorite icon". Typically this icon is displayed in the browser's address bar, or in front of the page title on the tab if the browser support tabs.

---

[2]The backslash before : is due to a technical issue: we want to prevent Pandoc from translating the link to HTML code `<a href="..."></a>`. More details at `https://github.com/jgm/pandoc/issues/2139`.

Below we show some sample YAML metadata (again, please note that these are *top-level* options):

```
---
title: "An Awesome Book"
author: "John Smith"
description: "This book introduces the ABC theory, and ..."
url: "https\://bookdown.org/john/awesome/"
github-repo: "john/awesome"
cover-image: "images/cover.png"
apple-touch-icon: "touch-icon.png"
apple-touch-icon-size: 120
favicon: "favicon.ico"
---
```

A nice effect of setting `description` and `cover-image` is that when you share the link of your book on some social network websites such as Twitter, the link can be automatically expanded to a card with the cover image and description of the book.

3.1.2   Bootstrap style

If you have used R Markdown before, you should be familiar with the Bootstrap style (http://getbootstrap.com), which is the default style of the HTML output of R Markdown. The output format function in **rmarkdown** is `html_document()`, and we have a corresponding format `html_book()` in **bookdown** using `html_document()` as the base format. In fact, there is a more general format `html_chapters()` in **bookdown** and `html_book()` is just its special case:

```
html_chapters(toc = TRUE, number_sections = TRUE, fig_caption = TRUE,
  lib_dir = "libs", template = bookdown_file("templates/default.html"),
  ..., base_format = rmarkdown::html_document, split_bib = TRUE,
  page_builder = build_chapter, split_by = c("section+number",
    "section", "chapter+number", "chapter", "rmd", "none"))
```

Note that it has a `base_format` argument that takes a base output format function, and `html_book()` is basically `html_chapters(base_format =` `rmarkdown::html_document)`. All arguments of `html_book()` are passed to `html_chapters()`:

```
html_book(...)
```

That means that you can use most arguments of `rmarkdown::html_document`, such as `toc` (whether to show the table of contents), `number_sections` (whether to number section headings), and so on. Again, check the help page of `rmarkdown::html_document` to see the full list of possible options. Note that the argument `self_contained` is hard-coded to `FALSE` internally, so you cannot change the value of this argument. We have explained the argument `split_by` in the previous section.

The arguments `template` and `page_builder` are for advanced users, and you do not need to understand them unless you have strong need to customize the HTML output, and those many options provided by `rmarkdown::html_document()` still do not give you what you want.

If you want to pass a different HTML template to the `template` argument, the template must contain three pairs of HTML comments, and each comment must be on a separate line:

- `<!--bookdown:title:start-->` and `<!--bookdown:title:end-->` to mark the title section of the book. This section will be placed only on the first page of the rendered book;
- `<!--bookdown:toc:start-->` and `<!--bookdown:toc:end-->` to mark the table of contents section, which will be placed on all HTML pages;
- `<!--bookdown:body:start-->` and `<!--bookdown:body:end-->` to mark the HTML body of the book, and the HTML body will be split into multiple separate pages. Recall that we merge all R Markdown or Markdown files, render them into a single HTML file, and split it.

You may open the default HTML template to see where these comments were inserted:

```
bookdown:::bookdown_file("templates/default.html")
# you may use file.edit() to open this file
```

Once you know how **bookdown** works internally to generate multiple-page HTML output, it will be easier to understand the argument page_builder, which is a function to compose each individual HTML page using the HTML fragments extracted from the above comment tokens. The default value of page_builder is a function build_chapter in **bookdown**, and its source code is relatively simple (ignore those internal functions like button_link()):

```
build_chapter = function(
  head, toc, chapter, link_prev, link_next, rmd_cur, html_cur, foot
) {
  # add a has-sub class to the <li> items that has sub lists
  toc = gsub('^(<li>)(.+<ul>)$', '<li class="has-sub">\\2', toc)
  paste(c(
    head,
    '<div class="row">',
    '<div class="col-sm-12">',
    toc,
    '</div>',
    '</div>',
    '<div class="row">',
    '<div class="col-sm-12">',
    chapter,
    '<p style="text-align: center;">',
    button_link(link_prev, 'Previous'),
    edit_link(rmd_cur),
    button_link(link_next, 'Next'),
    '</p>',
    '</div>',
    '</div>',
    foot
  ), collapse = '\n')
}
```

Basically, this function takes a number of components like the HTML head, the table of contents, the chapter body, and so on, and it is expected to return a character string which is the HTML source of a complete HTML page.

You may manipulate all components in this function using text-processing functions like `gsub()` and `paste()`.

What the default page builder does is to put TOC in the first row, the body in the second row, navigation buttons at the bottom of the body, and concatenate them with the HTML head and foot. Here is a sketch of the HTML source code that may help you understand the output of `build_chapter()`:

```html
<html>
  <head>
    <title>A Nice Book</title>
  </head>
  <body>

    <div class="row">TOC</div>

    <div class="row">
      CHAPTER BODY
      <p>
        <button>PREVIOUS</button>
        <button>NEXT</button>
      </p>
    </div>

  </body>
</html>
```

For all HTML pages, the main difference is the chapter body, and most of the rest of the elements are the same. The default output from `html_book()` will include the Bootstrap CSS and JavaScript files in the `<head>` tag.

The TOC is often used for navigation purposes. In the GitBook style, the TOC is displayed in the sidebar. For the Bootstrap style, we did not apply a special style to it, so it is shown as a plain unordered list (in the HTML tag `<ul>`). It is easy to turn this list into a navigation bar with some CSS techniques. We have provided a CSS file `toc.css` in this package that you can use, and you can find it here: `https://github.com/rstudio/bookdown/blob/master/inst/examples/css/toc.css`

You may copy this file to the root directory of your book, and apply it to the
HTML output via the `css` option, e.g.,

```
---
output:
  bookdown::html_book:
    toc: yes
    css: toc.css
---
```

There are many possible ways to turn `<ul>` lists into navigation menus if
you do a little bit searching on the web, and you can choose a menu style that
you like. The `toc.css` we just mentioned is a style with white menu texts on a
black background, and supports sub-menus (e.g., section titles are displayed
as drop-down menus under chapter titles).

As a matter of fact, you can get rid of the Bootstrap style in `html_document()`
if you set the `theme` option to `null`, and you are free to apply arbitrary styles
to the HTML output using the `css` option (and possibly the `includes` option
if you want to include arbitrary content in the HTML head/foot).

3.1.3   Tufte style

Like the Bootstrap style, the Tufte style is provided by an output format
`tufte_html_book()`, which is also a special case of `html_chapters()` using
`tufte::tufte_html()` as the base format. Please see the **tufte** package (Xie
and Allaire, 2016) if you are not familiar with the Tufte style. Basically, it is a
layout with a main column on the left and a margin column on the right. The
main body is in the main column, and the margin column is used to place
footnotes, margin notes, references, and margin figures, and so on.

All arguments of `tufte_html_book()` have exactly the same meanings as
`html_book()`, e.g., you can also customize the CSS via the `css` option. There
are a few elements that are specific to the Tufte style, though, such as margin
notes, margin figures, and full-width figures. These elements require special syntax to generate; please see the documentation of the **tufte** package.
Note that you do not need to do anything special to footnotes and references
(just use the normal Markdown syntax `^[footnote]` and `[@citation]`), since

they will be automatically put in the margin. A brief YAML example of the `tufte_html_book` format:

```
---
output:
  bookdown::tufte_html_book:
    toc: yes
    css: toc.css
---
```

3.2 LaTeX/PDF

We strongly recommend that you use an HTML output format instead of LaTeX when you develop a book, since you will not be too distracted by the typesetting details, which can bother you a lot if you constantly look at the PDF output of a book. Leave the job of careful typesetting to the very end (ideally after you have really finished the content of the book).

The LaTeX/PDF output format is provided by `pdf_book()` in **bookdown**. There is not a significant difference between `pdf_book()` and the `pdf_document()` format in **rmarkdown**. The main purpose of `pdf_book()` is to resolve the labels and cross-references written using the syntax described in Sections 2.4, 2.5, and 2.6. If the only output format that you want for a book is LaTeX/PDF, you may use the syntax specific to LaTeX, such as `\label{}` to label figures/tables/sections, and `\ref{}` to cross-reference them via their labels, because Pandoc supports LaTeX commands in Markdown. However, the LaTeX syntax is not portable to other output formats, such as HTML and e-books. That is why we introduced the syntax (`\#label`) for labels and `\@ref(label)` for cross-references.

There are some top-level YAML options that will be applied to the LaTeX output. For a book, you may change the default document class to `book` (the default is `article`), and specify a bibliography style required by your publisher. A brief YAML example:

```
---
documentclass: book
bibliography: [book.bib, packages.bib]
biblio-style: apalike
---
```

There are a large number of other YAML options that you can specify for LaTeX output, such as the paper size, font size, page margin, line spacing, font families, and so on. See `http://pandoc.org/MANUAL.html#variables-for-latex` for a full list of options.

The `pdf_book()` format is a general format like `html_book()`, and it also has a `base_format` argument:

```
pdf_book(toc = TRUE, number_sections = TRUE, fig_caption = TRUE,
    ..., base_format = rmarkdown::pdf_document, toc_unnumbered = TRUE,
    toc_appendix = FALSE, toc_bib = FALSE, quote_footer = NULL,
    highlight_bw = FALSE)
```

You can change the `base_format` function to other output format functions, and **bookdown** has provided a simple wrapper function `tufte_book2()`, which is basically `pdf_book(base_format = tufte::tufte_book)`, to produce a PDF book using the Tufte PDF style (again, see the **tufte** package).

3.3   E-Books

Currently **bookdown** provides two e-book formats, EPUB and MOBI. Books in these formats can be read on devices like smartphones, tablets, or special e-readers such as Kindle.

3.3.1 EPUB

To create an EPUB book, you can use the `epub_book()` format. It has some options in common with `rmarkdown::html_document()`:

```
epub_book(fig_width = 5, fig_height = 4, dev = "png",
  fig_caption = TRUE, number_sections = TRUE, toc = FALSE,
  toc_depth = 3, stylesheet = NULL, cover_image = NULL,
  metadata = NULL, chapter_level = 1, epub_version = c("epub3",
    "epub"), md_extensions = NULL, pandoc_args = NULL)
```

The option `toc` is turned off because the e-book reader can often figure out a TOC automatically from the book, so it is not necessary to add a few pages for the TOC. There are a few options specific to EPUB:

- `stylesheet`: It is similar to the `css` option in HTML output formats, and you can customize the appearance of elements using CSS.
- `cover_image`: The path to the cover image of the book.
- `metadata`: The path to an XML file for the metadata of the book (see Pandoc documentation for more details).
- `chapter_level`: Internally an EPUB book is a series of "chapter" files, and this option determines the level by which the book is split into these files. This is similar to the `split_by` argument of HTML output formats we mentioned in Section 3.1, but an EPUB book is a single file, and you will not see these "chapter" files directly. The default level is the first level, and if you set it to 2, it means the book will be organized by section files internally, which may allow the reader to load the book more quickly.
- `epub_version`: Version 3 or 2 of EPUB.

An EPUB book is essentially a collection of HTML pages, e.g., you can apply CSS rules to its elements, embed images, insert math expressions (because MathML is partially supported), and so on. Figure/table captions, cross-references, custom blocks, and citations mentioned in Chapter 2 also work for EPUB. You may compare the EPUB output of this book to the HTML output, and you will see that the only major difference is the visual appearance.

There are several EPUB readers available, including Calibre (https://www.calibre-ebook.com), Apple's iBooks, and Google Play Books.

3.3.2 MOBI

MOBI e-books can be read on Amazon's Kindle devices. Pandoc does not support MOBI output natively, but Amazon has provided a tool named Kindle-Gen (https://www.amazon.com/gp/feature.html?docId=1000765211) to create MOBI books from other formats, including EPUB and HTML. We have provided a simple wrapper function `kindlegen()` in **bookdown** to call Kindle-Gen to convert an EPUB book to MOBI. This requires you to download KindleGen first, and make sure the KindleGen executable can be found via the system environment variable PATH.

Another tool to convert EPUB to MOBI is provided by Calibre. Unlike Kindle-Gen, Calibre is open-source and free, and supports conversion among many more formats. For example, you can convert HTML to EPUB, Word documents to MOBI, and so on. The function `calibre()` in **bookdown** is a wrapper function of the command-line utility `ebook-convert` in Calibre. Similarly, you need to make sure that the executable `ebook-convert` can be found via the environment variable PATH. If you use OS X, you can install both KindleGen and Calibre via Homebrew-Cask (https://caskroom.github.io), so you do not need to worry about the PATH issue.

3.4 A single document

Sometimes you may not want to write a book, but a single long-form article or report instead. Usually what you do is call `rmarkdown::render()` with a certain output format. The main features missing there are the automatic numbering of figures/tables/equations, and cross-referencing figures/tables/equations/sections. We have factored out these features from **bookdown**, so that you can use them without having to prepare a book of multiple Rmd files.

The functions `html_document2()`, `tufte_html2()`, `pdf_document2()`, `word_document2()`, `tufte_handout2()`, and `tufte_book2()` are designed for this purpose. If you render an R Markdown document with the output format, say, `bookdown::html_document2`, you will get figure/table numbers and be able to cross-reference them in the single HTML page using the syntax described in Chapter 2.

The above HTML and PDF output format functions are basically wrappers of output formats `bookdown::html_book` and `bookdown::pdf_book`, in the sense that they changed the `base_format` argument. For example, you can take a look at the source code of `pdf_document2`:

```
bookdown::pdf_document2
```

```
## function(...) {
##    pdf_book(..., base_format = rmarkdown::pdf_document)
## }
## <environment: namespace:bookdown>
```

After you know this fact, you can apply the same idea to other output formats by using the appropriate `base_format`. For example, you can port the **bookdown** features to the `jss_article` format in the **rticles** package (Allaire et al., 2016b) by using the YAML metadata:

```
output:
  bookdown::pdf_book:
    base_format: rticles::jss_article
```

Then you will be able to use all features we introduced in Chapter 2.

Although the `gitbook()` format was designed primarily for books, you can actually also apply it to a single R Markdown document. The only difference is that there will be no search button on the single page output, because you can simply use the searching tool of your web browser to find text (e.g., press `Ctrl + F` or `Command + F`). You may also want to set the option `split_by` to `none` to only generate a single output page, in which case there will not be any navigation buttons, since there are no other pages to navigate to. You can still generate multiple-page HTML files if you like. Another option you may want to use is `self_contained = TRUE` when it is only a single output page.

4

Customization

As we mentioned in the very beginning of this book, you are expected to have some basic knowledge about R Markdown, and we have been focusing on introducing the **bookdown** features instead of **rmarkdown**. In fact, R Markdown is highly customizable, and there are many options that you can use to customize the output document. Depending on how much you want to customize the output, you may use some simple options in the YAML metadata, or just replace the entire Pandoc template.

4.1 YAML options

For most types of output formats, you can customize the syntax highlighting styles using the `highlight` option of the specific format. Currently, the possible styles are `default`, `tango`, `pygments`, `kate`, `monochrome`, `espresso`, `zenburn`, and `haddock`. For example, you can choose the `tango` style for the `gitbook` format:

```
---
output:
  bookdown::gitbook:
    highlight: tango
---
```

For HTML output formats, you are most likely to use the `css` option to provide your own CSS stylesheets to customize the appearance of HTML elements. There is an option `includes` that applies to more formats, including HTML and LaTeX. The `includes` option allows you to insert arbitrary custom content before and/or after the body of the output. It has three sub-options: `in_header`, `before_body`, and `after_body`. You need to know the basic structure of an HTML or LaTeX document to understand these options. The source of an HTML document looks like this:

```
<html>

  <head>
  <!-- head content here, e.g. CSS and JS -->
  </head>

  <body>
  <!-- body content here -->
  </body>

</html>
```

The `in_header` option takes a file path and inserts it into the `<head>` tag. The `before_body` file will be inserted right below the opening `<body>` tag, and `after_body` is inserted before the closing tag `</body>`.

A LaTeX source document has a similar structure:

```
\documentclass{book}

% LaTeX preamble
% insert in_header here

\begin{document}
% insert before_body here

% body content here

% insert after_body here
\end{document}
```

The `includes` option is very useful and flexible. For HTML output, it means you can insert arbitrary HTML code into the output. For example, when you have LaTeX math expressions rendered via the MathJax library in the HTML output, and want the equation numbers to be displayed on the left (default is on the right), you can create a text file that contains the following code:

```
<script type="text/x-mathjax-config">
MathJax.Hub.Config({
  TeX: { TagSide: "left" }
});
</script>
```

Let's assume the file is named `mathjax-number.html`, and it is in the root directory of your book (the directory that contains all your Rmd files). You can insert this file into the HTML head via the `in_header` option, e.g.,

```
---

output:
  bookdown::gitbook:
    includes:
      in_header: mathjax-number.html

---
```

Another example is to enable comments or discussions on your HTML pages. There are several possibilities, such as Disqus (`https://disqus.com`) or Hypothesis (`https://hypothes.is`). These services can be easily embedded in your HTML book via the `includes` option (see Section 5.5 for details).

Similarly, if you are familiar with LaTeX, you can add arbitrary LaTeX code to the preamble. That means you can use any LaTeX packages and set up any package options for your book. For example, this book used the `in_header` option to use a few more LaTeX packages like **booktabs** (for better-looking tables) and **longtable** (for tables that span across multiple pages), and applied a fix to an XeLaTeX problem that links on graphics do not work:

```
\usepackage{booktabs}
\usepackage{longtable}

\ifxetex
```

```
\usepackage{letltxmacro}
\setlength{\XeTeXLinkMargin}{1pt}
\LetLtxMacro\SavedIncludeGraphics\includegraphics
\def\includegraphics#1#{% #1 catches optional stuff (star/opt. arg.)
  \IncludeGraphicsAux{#1}%
}%
\newcommand*{\IncludeGraphicsAux}[2]{%
  \XeTeXLinkBox{%
    \SavedIncludeGraphics#1{#2}%
  }%
}%
\fi
```

The above LaTeX code is saved in a file `preamble.tex`, and the YAML metadata looks like this:

```
---
output:
  bookdown::pdf_book:
    includes:
      in_header: preamble.tex
---
```

4.2   Theming

Sometimes you may want to change the overall theme of the output, and usually this can be done through the `in_header` option described in the previous
section, or the `css` option if the output is HTML. Some output formats have
their unique themes, such as `gitbook`, `tufte_html_book`, and `tufte_book2`,
and you may not want to customize these themes too much. By comparison,
the output formats `html_book()` and `pdf_book()` are not tied to particular
themes and more customizable.

As mentioned in Section 3.1.2, the default style for `html_book()` is the Bootstrap style. The Bootstrap style actually has several built-in themes that you can use, including `default`, `cerulean`, `journal`, `flatly`, `readable`, `spacelab`, `united`, `cosmo`, `lumen`, `paper`, `sandstone`, `simplex`, and `yeti`. You can set the theme via the `theme` option, e.g.,

```
---
output:
  bookdown::html_book:
    theme: united
---
```

If you do not like any of these Bootstrap styles, you can set `theme` to `null`, and apply your own CSS through the `css` or `includes` option.

For `pdf_book()`, besides the `in_header` option mentioned in the previous section, another possibility is to change the document class. There are many possible LaTeX classes for books, such as **memoir** (`https://www.ctan.org/pkg/memoir`), **amsbook** (`https://www.ctan.org/pkg/amsbook`), KOMA-Script (`https://www.ctan.org/pkg/koma-script`) and so on. Here is a brief sample of the YAML metadata specifying the `scrbook` class from the KOMA-Script package:

```
---
documentclass: scrbook
output:
  bookdown::pdf_book:
    template: null
---
```

Some publishers (e.g., Springer and Chapman & Hall/CRC) have their own LaTeX style or class files. You may try to change the `documentclass` option to use their document classes, although typically it is not as simple as that. You may end up using `in_header`, or even design a custom Pandoc LaTeX template to accommodate these document classes.

Note that when you change `documentclass`, you are likely to specify an additional Pandoc argument `--chapters` so that Pandoc knows the first-level headers should be treated as chapters instead of sections (this is the default when documentclass is book), e.g.,

```
documentclass: krantz
output:
  bookdown::pdf_book:
    pandoc_args: --chapters
```

4.3 Templates

When Pandoc converts Markdown to another output format, it uses a template under the hood. The template is a plain-text file that contains some variables of the form $variable$. These variables will be replaced by their values generated by Pandoc. Below is a very brief template for HTML output:

```
<html>
  <head>
    <title>$title$</title>
  </head>

  <body>
  $body$
  </body>
</html>
```

It has two variables title and body. The value of title comes from the title field of the YAML metadata, and body is the HTML code generated from the body of the Markdown input document. For example, suppose we have a Markdown document:

```
---

title: A Nice Book
```

```
---

# Introduction

This is a **nice** book!
```

If we use the above template to generate an HTML document, its source code will be like this:

```html
<html>
  <head>
    <title>A Nice Book</title>
  </head>

  <body>

  <h1>Introduction</h1>

  <p>This is a <strong>nice</strong> book!</p>

  </body>
</html>
```

The actual HTML, LaTeX, and EPUB templates are more complicated, but the idea is the same. You need to know what variables are available: some variables are built-in Pandoc variables, and some can be either defined by users in the YAML metadata, or passed from the command-line option -v or --variable. Some variables only make sense in specific output formats, e.g., the documentclass variable is only used in LaTeX output. Please see the documentation of Pandoc to learn more about these variables, and you can find all default Pandoc templates in the GitHub repository https://github.com/jgm/pandoc-templates.

Note that for HTML output, **bookdown** requires some additional comment tokens in the template, and we have explained them in Section 3.1.2.

4.4 Configuration

We have mentioned `rmd_files` in Section 1.3, and there are more (optional) settings you can configure for a book in `_bookdown.yml`:

- `book_filename`: the filename of the main Rmd file, i.e., the Rmd file that is merged from all chapters; by default, it is named `_main.Rmd`.
- `before_chapter_script`: one or multiple R scripts to be executed before each chapter, e.g., you may want to clear the workspace before compiling each chapter, in which case you can use `rm(list = ls(all = TRUE))` in the R script.
- `after_chapter_script`: similar to `before_chapter_script`, and the R script is executed after each chapter.
- `edit`: a link that collaborators can click to edit the Rmd source document of the current page; this was designed primarily for GitHub repositories, since it is easy to edit arbitrary plain-text files on GitHub even in other people's repositories (if you do not have write access to the repository, GitHub will automatically fork it and let you submit a pull request after you finish editing the file). This link should have %s in it, which will be substituted by the actual Rmd filename for each page.
- `rmd_subdir`: whether to search for book source Rmd files in subdirectories (by default, only the root directory is searched).
- `output_dir`: the output directory of the book (`_book` by default); this setting is read and used by `render_book()`.
- `clean`: a vector of files and directories to be cleaned by the `clean_book()` function.

Here is a sample `_bookdown.yml`:

```
book_filename: "my-book.Rmd"
before_chapter_script: ["script1.R", "script2.R"]
after_chapter_script: "script3.R"
edit: https://github.com/rstudio/bookdown-demo/edit/master/%s
output_dir: "book-output"
clean: ["my-book.bbl", "R-packages.bib"]
```

4.5 Internationalization

If the language of your book is not English, you will need to translate certain English words and phrases into your language, such as the words "Figure" and "Table" when figures/tables are automatically numbered in the HTML output. Internationalization may not be an issue for LaTeX output, since some LaTeX packages can automatically translate these terms into the local language, such as the **ctexcap** package for Chinese.

For non-LaTeX output, you can set the language field in the configuration file _bookdown.yml. Currently the default settings are:

```
language:
  label:
    fig: 'Figure '
    tab: 'Table '
    eq: 'Equation '
    thm: 'Theorem '
    lem: 'Lemma '
    def: 'Definition '
    cor: 'Corollary '
    prp: 'Proposition '
    ex: 'Example '
    proof: 'Proof. '
    remark: 'Remark. '
  ui:
    edit: Edit
    chapter_name: ''
```

For example, if you want FIGURE x.x instead of Figure x.x, you can change fig to "FIGURE ":

```
language:
  label:
    fig: "FIGURE "
```

The fields under `ui` are used to specify some terms in the user interface. The `edit` field specifies the text associated with the `edit` link in `_bookdown.yml` (Section 4.4). The `chapter_name` field can be either a character string to be prepended to chapter numbers in chapter titles (e.g., `'CHAPTER '`), or an R function that takes the chapter number as the input and returns a string as the new chapter number (e.g., `!expr function(i) paste('Chapter', i)`). If it is a character vector of length 2, the chapter title prefix will be `paste0(chapter_name[1], i, chapter_name[2])`, where `i` is the chapter number.

There is one caveat when you write in a language that uses multibyte characters, such as Chinese, Japanese, and Korean (CJK): Pandoc cannot generate identifiers from section headings that are pure CJK characters, so you will not be able to cross-reference sections (they do not have labels), unless you manually assign identifiers to them by appending `{#identifier}` to the section heading, where `identifier` is an identifier of your choice.

5

Editing

In this chapter, we explain how to edit, build, preview, and serve the book locally. You can use any text editors to edit the book, and we will show some tips for using the RStudio IDE. We will introduce the underlying R functions for building, previewing, and serving the book before we introduce the editor, so that you really understand what happens behind the scenes when you click a certain button in the RStudio IDE, and can also customize other editors calling these functions.

5.1 Build the book

To build all Rmd files into a book, you can call the `render_book()` function in **bookdown**. Below are the arguments of `render_book()`:

```
render_book(input, output_format = NULL, ..., clean = TRUE,
  envir = parent.frame(), clean_envir = !interactive(),
  output_dir = NULL, new_session = NA, preview = FALSE,
  encoding = "UTF-8")
```

The most important argument is `output_format`, which can take a character string of the output format (e.g., `'bookdown::gitbook'`). You can leave this argument empty, and the default output format will be the first output format specified in the YAML metadata of the first Rmd file or a separate YAML file _output.yml, as mentioned in Section 4.4. If you plan to generate multiple output formats for a book, you are recommended to specify all formats in _output.yml.

Once all formats are specified in _output.yml, it is easy to write an R or Shell

script or Makefile to compile the book. Below is a simple example of using a
Shell script to compile a book to HTML (with the GitBook style) and PDF:

```
#!/usr/bin/env Rscript

bookdown::render_book("index.Rmd", "bookdown::gitbook")
bookdown::render_book("index.Rmd", "bookdown::pdf_book")
```

The Shell script does not work on Windows (not strictly true, though), but
hopefully you get the idea.

The argument . . . is passed to the output format function. Arguments clean
and envir are passed to rmarkdown::render(), to decide whether to clean
up the intermediate files, and specify the environment to evaluate R code,
respectively.

The output directory of the book can be specified via the output_dir argument. By default, the book is generated to the _book directory. This can also
be changed via the output_dir field in the configuration file _bookdown.yml,
so that you do not have to specify it multiple times for rendering a book to
multiple output formats. The new_session argument has been explained in
Section 1.4. When you set preview = TRUE, only the Rmd files specified in
the input argument are rendered, which can be convenient when previewing a certain chapter, since you do not recompile the whole book, but when
publishing a book, this argument should certainly be set to FALSE.

When you render the book to multiple formats in the same R session,
you need to be careful because the next format may have access to R objects created from the previous format. You are recommended to render
the book with a clean environment for each output format. The argument
clean_envir can be used to clean all objects in the environment specified
by envir. By default, it is TRUE for non-interactive R sessions (e.g., in batch
mode). Note that even clean_envir = TRUE does not really guarantee the R
session is clean. For example, packages loaded when rendering the previous
format will remain in the session for the next output format. To make sure
each format is rendered in a completely clean R session, you have to actually
launch a new R session to build each format, e.g., use the command line

```
Rscript -e "bookdown::render_book('index.Rmd', 'bookdown::gitbook')"
Rscript -e "bookdown::render_book('index.Rmd', 'bookdown::pdf_book')"
```

A number of output files will be generated by `render_book()`. Sometimes you may want to clean up the book directory and start all over again, e.g., remove the figure and cache files that were generated automatically from **knitr**. The function `clean_book()` was designed for this purpose. By default, it tells you which output files you can possibly delete. If you have looked at this list of files, and are sure no files were mistakenly identified as output files (you certainly do not want to delete an input file that you created by hand), you can delete all of them using `bookdown::clean_book(TRUE)`. Since deleting files is a relatively dangerous operation, we would recommend that you maintain your book through version control tools such as GIT, or a service that supports backup and restoration, so you will not lose certain files forever if you delete them by mistake.

5.2 Preview a chapter

Building the whole book can be slow when the size of the book is big. Two things can affect the speed of building a book: the computation in R code chunks, and the conversion from Markdown to other formats via Pandoc. The former can be improved by enabling caching in **knitr** using the chunk option `cache = TRUE`, and there is not much you can do to make the latter faster. However, you can choose to render only one chapter at a time using the function `preview_chapter()` in **bookdown**, and usually this will be much faster than rendering the whole book. Only the Rmd files passed to `preview_chapter()` will be rendered.

Previewing the current chapter is helpful when you are only focusing on that chapter, since you can quickly see the actual output as you add more content or revise the chapter. Although the preview works for all output formats, we recommend that you preview the HTML output.

One downside of previewing a chapter is that the cross-references to other chapters will not work, since **bookdown** knows nothing about other chapters in this case. That is a reasonably small price to pay for the gain in speed. Since previewing a chapter only renders the output for that specific chapter, you should not expect that the content of other chapters is correctly rendered as well. For example, when you navigate to a different chapter, you are actually viewing the old output of that chapter (which may not even exist).

5.3   Serve the book

Instead of running `render_book()` or `preview_chapter()` over and over again, you can actually live preview the book in the web browser, and the only thing you need to do is save the Rmd file. The function `serve_book()` in **bookdown** can start a local web server to serve the HTML output based on the **servr** package (Xie, 2016d).

```
serve_book(dir = ".", output_dir = "_book", preview = TRUE,
  in_session = TRUE, daemon = FALSE, ...)
```

You pass the root directory of the book to the `dir` argument, and this function will start a local web server so you can view the book output using the server. The default URL to access the book output is `http://127.0.0.1:4321`. If you run this function in an interactive R session, this URL will be automatically opened in your web browser. If you are in the RStudio IDE, the RStudio Viewer will be used as the default web browser, so you will be able to write the Rmd source files and preview the output in the same environment (e.g, source on the left and output on the right).

The server will listen to changes in the book root directory: whenever you modify any files in the book directory, `serve_book()` can detect the changes, recompile the Rmd files, and refresh the web browser automatically. If the modified files do not include Rmd files, it just refreshes the browser (e.g., if you only updated a certain CSS file). This means once the server is launched, all you have to do next is simply write the book and save the files. Compilation and preview will take place automatically as you save files.

If it does not really take too much time to recompile the whole book, you may set the argument `preview = FALSE`, so that every time you update the book, the whole book is recompiled, otherwise only the modified chapters are recompiled via `preview_chapter()`.

The arguments `daemon` and ... are passed to `servr::httw()`, and please see its help page to see all possible options, such as `port`. There are pros and cons of using `in_session = TRUE` or `FALSE`:

- For `in_session = TRUE`, you will have access to all objects created in the book in the current R session: if you use a daemonized server (via the argument `daemon = TRUE`), you can check the objects at any time when the current R session is not busy; otherwise you will have to stop the server before you can check the objects. This can be useful when you need to interactively explore the R objects in the book. The downside of `in_session = TRUE` is that the output may be different with the book compiled from a fresh R session, because the state of the current R session may not be clean.
- For `in_session = FALSE`, you do not have access to objects in the book from the current R session, but the output is more likely to be reproducible since everything is created from new R sessions. Since this function is only for previewing purposes, the cleanness of the R session may not be a big concern.

You may choose `in_session = TRUE` or `FALSE` depending on your specific use cases. Eventually, you should run `render_book()` from a fresh R session to generate a reliable copy of the book output.

5.4 RStudio IDE

We recommend that you upgrade[1] your RStudio IDE if your version is lower than 1.0.0. As mentioned in Section 1.3, all R Markdown files must be encoded in UTF-8. This is important especially when your files contain multi-byte characters. To save a file with the UTF-8 encoding, you can use the menu `File -> Save with Encoding`, and choose `UTF-8`.

---

[1]`https://www.rstudio.com/products/rstudio/download/`

When you click the `Knit` button to compile an R Markdown document in the RStudio IDE, the default function called by RStudio is `rmarkdown::render()`, which is not what we want for books. To call the function `bookdown::render_book()` instead, you can set the `site` field to be `bookdown::bookdown_site` in the YAML metadata of the R Markdown document `index.Rmd`, e.g.,

```
---
title: "A Nice Book"
site: bookdown::bookdown_site
output:
  bookdown::gitbook: default
---
```

When you have set `site: bookdown::bookdown_site` in `index.Rmd`, RStudio will be able to discover the directory as a book source directory,[2] and you will see a button `Build Book` in the `Build` pane. You can click the button to build the whole book in different formats, and if you click the `Knit` button on the toolbar, RStudio will automatically preview the current chapter, and you do not need to use `preview_chapter()` explicitly.

The **bookdown** package comes with a few addins for RStudio. If you are not familiar with RStudio addins, you may check out the documentation at http://rstudio.github.io/rstudioaddins/. After you have installed the **bookdown** package and use RStudio v0.99.878 or later, you will see a drop-down menu on the toolbar named "Addins" and menu items like "Preview Book" and "Input LaTeX Math" after you open the menu.

The addin "Preview Book" calls `bookdown::serve_book()` to compile and serve the book. It will block your current R session, i.e., when `serve_book()` is running, you will not be able to do anything in the R console anymore. To avoid blocking the R session, you can daemonize the server using `bookdown::serve_book(daemon = TRUE)`. Note that this addin must be used when the current document opened in RStudio is under the root directory of your book, otherwise `serve_book()` may not be able to find the book source.

The addin "Input LaTeX Math" is essentially a small Shiny application that

---

[2]This directory has to be an RStudio project.

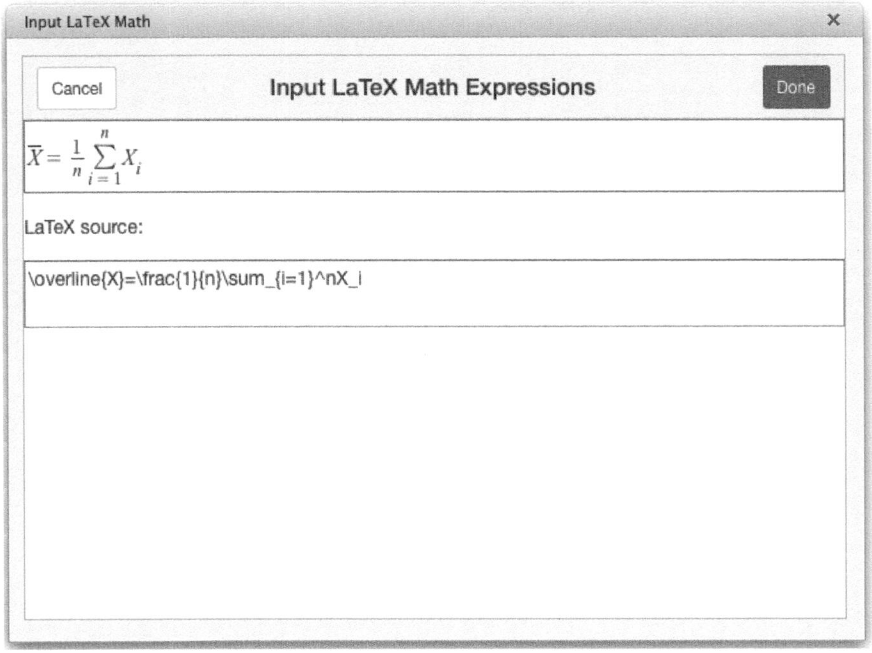

FIGURE 5.1: The RStudio addin to help input LaTeX math.

provides a text box to help you type LaTeX math expressions (Figure 5.1). As you type, you will see the preview of the math expression and its LaTeX source code. This will make it much less error-prone to type math expressions — when you type a long LaTeX math expression without preview, it is easy to make mistakes such as x_ij when you meant x_{ij}, or omitting a closing bracket. If you have selected a LaTeX math expression in the RStudio editor before clicking the addin, the expression will be automatically loaded and rendered in the text box. This addin was built on top of the MathQuill library (http://mathquill.com). It is not meant to provide full support to all LaTeX commands for math expressions, but should help you type some common math expressions.

There are also other R packages that provide addins to help you author books. The **citr** package (Aust, 2016) provides an addin named "Insert citations", which makes it easy to insert citations into R Markdown documents. It scans your bibliography databases, and shows all citation items in a drop-down menu, so you can choose from the list without remembering which citation key corresponds to which citation item (Figure 5.2).

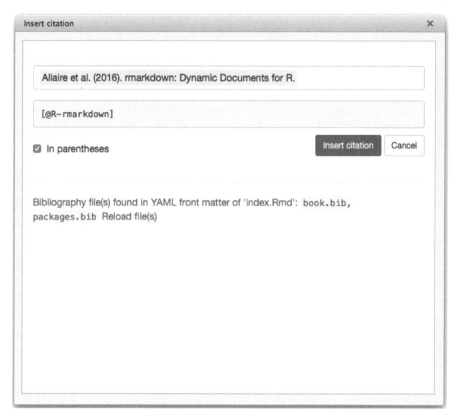

FIGURE 5.2: The RStudio addin to help insert citations.

5.5   Collaboration

Writing a book will almost surely involve more than a single person. You may have co-authors, and readers who give you feedback from time to time.

Since all book chapters are plain-text files, they are perfect for version control tools, which means if all your co-authors and collaborators have basic knowledge of a version control tool like GIT, you can collaborate with them on the book content using these tools. In fact, collaboration with GIT is possible even if they do not know how to use GIT, because GitHub has made it possible to create and edit files online right in your web browser. Only one person has to be familiar with GIT, and that person can set up the book repository. The rest of the collaborators can contribute content online, although they will have more freedom if they know the basic usage of GIT to work locally.

Readers can contribute in two ways. One way is to contribute content directly, and the easiest way, is through GitHub pull requests[3] if your book source is hosted on GitHub. Basically, any GitHub user can click the edit button on the page of an Rmd source file, edit the content, and submit the changes to you for your approval. If you are satisfied with the changes proposed (you can clearly see what exactly was changed), you can click a "Merge" button to merge the changes. If you are not satisfied, you can provide your feedback in the pull request, so the reader can further revise it according to your requirements. We mentioned the edit button in the GitBook style in Section 3.1.1. That button is linked to the Rmd source of each page, and can guide you to create the pull request. There is no need to write emails back and forth to communicate simple changes, such as fixing a typo.

Another way for readers to contribute to your book is to leave comments. Comments can be left in multiple forms: emails, GitHub issues, or HTML page comments. Here we use Disqus (see Section 4.1) as an example. Disqus is a service to embed a discussion area on your web pages, and can be loaded via JavaScript. You can find the JavaScript code after you register and create a new forum on Disqus, which looks like this:

```html
<div id="disqus_thread"></div>
<script>
(function() { // DON'T EDIT BELOW THIS LINE
var d = document, s = d.createElement('script');
s.src = '//yihui.disqus.com/embed.js';
s.setAttribute('data-timestamp', +new Date());
(d.head || d.body).appendChild(s);
})();
</script>
<noscript>Please enable JavaScript to view the
<a href="https://disqus.com/?ref_noscript">
  comments powered by Disqus.</a></noscript>
```

---

[3]https://help.github.com/articles/about-pull-requests/

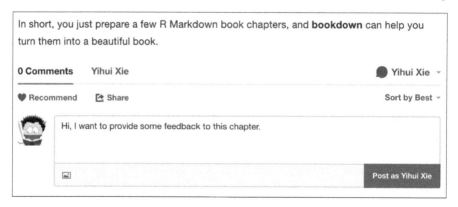

FIGURE 5.3: A book page with a discussion area.

Note that you will need to replace the name yihui with your own forum name (this name has to be provided when you create a new Disqus forum). You can save the code to an HTML file named, for example, disqus.html. Then you can embed it at the end of every page via the after_body option (Figure 5.3 shows what the discussion area looks like):

```
---
output:
  bookdown::gitbook:
    includes:
      after_body: disqus.html
---
```

6

Publishing

As you develop the book, you make the draft book available to the public to get early feedback from readers, e.g., publish it to a website. After you finish writing the book, you need to think about options to formally publish it as either printed copies or e-books.

6.1 RStudio Connect

In theory, you can render the book by yourself and publish the output anywhere you want. For example, you can host the HTML files on your own web server. We have provided a function `publish_book()` in **bookdown** to make it very simple to upload your book to `https://bookdown.org`, which is a website provided by RStudio to host your books for free. This website is built on top of "RStudio Connect",[1] an RStudio product that allows you to deploy a variety of R-related applications to a server, including R Markdown documents, Shiny applications, R plots, and so on.

You do not have to know much about RStudio Connect to publish your book to bookdown.org. Basically you sign up at `https://bookdown.org/connect/`, and the first time you try to run `bookdown::publish_book()`, you will be asked to authorize **bookdown** to publish to your bookdown.org account. In the future, you simply call `publish_book()` again and **bookdown** will no longer ask for anything.

```
publish_book(name = NULL, account = NULL, server = NULL,
  render = c("none", "local", "server"))
```

---

[1] https://www.rstudio.com/products/connect/

The only argument of `publish_book()` that you may want to touch is `render`. It determines whether you want to render the book before publishing. If you have run `render_book()` before, you do not need to change this argument, otherwise you may set it to `'local'`:

```
bookdown::publish_book(render = "local")
```

If you have set up your own RStudio Connect server, you can certainly publish the book to that server instead of bookdown.org.

6.2 GitHub

You can host your book on GitHub for free via GitHub Pages (`https://pages.github.com`). GitHub supports Jekyll (`http://jekyllrb.com`), a static website builder, to build a website from Markdown files. That may be the more common use case of GitHub Pages, but GitHub also supports arbitrary static HTML files, so you can just host the HTML output files of your book on GitHub.

One approach is to publish your book as a GitHub Pages site from a `/docs` folder on your `master` branch as described in GitHub Help.[2] First, set the output directory of your book to be `/docs` by adding the line `output_dir: "docs"` to the configuration file `_bookdown.yml`. Then, after pushing your changes to GitHub, go to your repository's settings and under "GitHub Pages" change the "Source" to be "master branch /docs folder".

An alternative approach is to create a `gh-pages` branch in your repository, build the book, put the HTML output (including all external resources like images, CSS, and JavaScript files) in this branch, and push the branch to the remote repository. If your book repository does not have the `gh-pages` branch, you may use the following commands to create one:

---

[2]`http://bit.ly/2cvloKV`

```
# assume you have initialized the git repository,
# and are under the directory of the book repository now

# create a branch named gh-pages and clean up everything
git checkout --orphan gh-pages
git rm -rf .

# create a hidden file .nojekyll
touch .nojekyll
git add .nojekyll

git commit -m"Initial commit"
git push origin gh-pages
```

The hidden file .nojekyll tells GitHub that your website is not to be built via Jekyll, since the **bookdown** HTML output is already a standalone website. If you are on Windows, you may not have the touch command, and you can create the file in R using file.create('.nojekyll').

After you have set up GIT, the rest of work can be automated via a script (Shell, R, or Makefile, depending on your preference). Basically, you compile the book to HTML, then run git commands to push the files to GitHub, but you probably do not want to do this over and over again manually and locally. It can be very handy to automate the publishing process completely on the cloud, so once it is set up correctly, all you have to do next is write the book and push the Rmd source files to GitHub, and your book will always be automatically built and published from the server side.

One service that you can utilize is Travis CI (https://travis-ci.org). It is free for public repositories on GitHub, and was designed for continuous integration (CI) of software packages. Travis CI can be connected to GitHub in the sense that whenever you push to GitHub, Travis can be triggered to run certain commands/scripts on the latest version of your repository.[3] These commands are specified in a YAML file named .travis.yml in the root di-

---

[3]You need to authorize the Travis CI service for your repository on GitHub first. See https://docs.travis-ci.com/user/getting-started/ for how to get started with Travis CI.

rectory of your repository, and they are usually for the purpose of testing software, but in fact they are quite open-ended, meaning that you can run arbitrary commands on a Travis (virtual) machine. That means you can certainly run your own scripts to build your book on Travis. Note that Travis only supports Ubuntu and Mac OS X at the moment, so you should have some basic knowledge about Linux/Unix commands.

The next question is, how to publish the book built on Travis to GitHub? Basically you have to grant Travis write access to your GitHub repository. This authorization can be done in several ways, and the easiest one to beginners may be a personal access token. Here are a few steps you may follow:

1. Create a personal access token[4] for your account on GitHub (make sure to enable the "repo" scope so that using this token will enable writing to your GitHub repos).
2. Encrypt it in the environment variable `GITHUB_PAT` via command line `travis encrypt` and store it in `.travis.yml`, e.g `travis encrypt GITHUB_PAT=TOKEN`. If you do not know how to install or use the Travis command-line tool, simply save this environment variable via `https://travis-ci.org/user/repo/settings` where `user` is your GitHub ID, and `repo` is the name of the repository.
3. You can clone this `gh-pages` branch on Travis using your GitHub token, add the HTML output files from R Markdown (do not forget to add figures and CSS style files as well), and push to the remote repository.

Assume you are in the `master` branch right now (where you put the Rmd source files), and have compiled the book to the `_book` directory. What you can do next on Travis is:

```
# configure your name and email if you have not done so
git config --global user.email "you@example.com"
git config --global user.name "Your Name"

# clone the repository to the book-output directory
git clone -b gh-pages \
```

---
[4]`http://bit.ly/2cEBYWB`

```
  https://${GITHUB_PAT}@github.com/${TRAVIS_REPO_SLUG}.git \
  book-output
cd book-output
cp -r ../_book/* ./
git add --all *
git commit -m"Update the book"
git push origin gh-pages
```

The variable name GITHUB_PAT and the directory name book-output are arbitrary, and you can use any names you prefer, as long as the names do not conflict with existing environment variable names or directory names. This script, together with the build script we mentioned in Section 5.1, can be put in the master branch as Shell scripts, e.g., you can name them as _build.sh and _deploy.sh. Then your .travis.yml may look like this:

```
language: r

env:
  global:
    - secure: A_LONG_ENCRYPTED_STRING

before_script:
  - chmod +x ./_build.sh
  - chmod +x ./_deploy.sh

script:
  - ./_build.sh
  - ./_deploy.sh
```

The language key tells Travis to use a virtual machine that has R installed. The secure key is your encrypted personal access token. If you have already saved the GITHUB_PAT variable using the web interface on Travis instead of the command-line tool travis encrypt, you can leave out this key.

Since this Travis service is primarily for checking R packages, you will also need a (fake) DESCRIPTION file as if the book repository were an R package. The only thing in this file that really matters is the specification of dependencies. All dependencies will be installed via the **devtools** package. If a dependency is on CRAN or BioConductor, you can simply list it in the Imports field of the DESCRIPTION file. If it is on GitHub, you may use the Remotes field to list its repository name. Below is an example:

```
Package: placeholder
Title: Does not matter.
Version: 0.0.1
Imports: bookdown, ggplot2
Remotes: rstudio/bookdown
```

If you use the container-based infrastructure[5] on Travis, you can enable caching by using sudo: false in .travis.yml. Normally you should cache at least two types of directories: the figure directory (e.g., _main_files) and the cache directory (e.g., _main_cache). These directory names may also be different if you have specified the **knitr** chunk options fig.path and cache.path, but I'd strongly recommend you not to change these options. The figure and cache directories are stored under the _bookdown_files directory of the book root directory. A .travis.yml file that has enabled caching of **knitr** figure and cache directories may have additional configurations sudo and cache like this:

```
sudo: false

cache:
  packages: yes
  directories:
    - $TRAVIS_BUILD_DIR/_bookdown_files
```

If your book is very time-consuming to build, you may use the above configurations on Travis to save time. Note that packages: yes means the R packages installed on Travis are also cached.

All above scripts and configurations can be found in the bookdown-demo repository: https://github.com/rstudio/bookdown-demo/. If you copy them to your own repository, please remember to change the secure key in .travis.yml using your own encrypted variable GITHUB_PAT.

---

[5]https://docs.travis-ci.com/user/workers/container-based-infrastructure/

GitHub and Travis CI are certainly not the only choices to build and publish your book. You are free to store and publish the book on your own server.

6.3 Publishers

Besides publishing your book online, you can certainly consider publishing it with a publisher. For example, this book was published with Chapman & Hall/CRC, and there is also a free online version at https://bookdown.org/ yihui/bookdown/ (with an agreement with the publisher). Another option that you can consider is self-publishing (https://en.wikipedia.org/wiki/ Self-publishing) if you do not want to work with an established publisher.

It will be much easier to publish a book written with **bookdown** if the publisher you choose supports LaTeX. For example, Chapman & Hall provides a LaTeX class named krantz.cls, and Springer provides svmono.cls. To apply these LaTeX classes to your PDF book, set documentclass in the YAML metadata of index.Rmd to the class filename (without the extension .cls).

The LaTeX class is the most important setting in the YAML metadata. It controls the overall style of the PDF book. There are often other settings you want to tweak, and we will show some details about this book below.

The YAML metadata of this book contains these settings:

```
documentclass: krantz
lot: yes
lof: yes
fontsize: 12pt
monofont: "Source Code Pro"
monofontoptions: "Scale=0.7"
```

The field lot: yes means we want the List of Tables, and similarly, lof means List of Figures. The base font size is 12pt, and we used Source Code

Pro[6] as the monospaced (fixed-width) font, which is applied to all program code in this book.

In the LaTeX preamble (Section 4.1), we have a few more settings. First, we set the main font to be Alegreya[7], and since this font does not have the SMALL CAPITALS feature, we used the Alegreya SC font.

```
\setmainfont[
  UprightFeatures={SmallCapsFont=AlegreyaSC-Regular}
]{Alegreya}
```

The following commands make floating environments less likely to float by allowing them to occupy larger fractions of pages without floating.

```
\renewcommand{\textfraction}{0.05}
\renewcommand{\topfraction}{0.8}
\renewcommand{\bottomfraction}{0.8}
\renewcommand{\floatpagefraction}{0.75}
```

Since `krantz.cls` provided an environment `VF` for quotes, we redefine the standard `quote` environment to `VF`. You can see its style in Section 2.1.

```
\renewenvironment{quote}{\begin{VF}}{\end{VF}}
```

Then we redefine hyperlinks to be footnotes, because when the book is printed on paper, readers are not able to click on links in text. Footnotes will tell them what the actual links are.

```
\let\oldhref\href
\renewcommand{\href}[2]{#2\footnote{\url{#1}}}
```

We also have some settings for the `bookdown::pdf_book` format in `_output.yml`:

---

[6]https://www.fontsquirrel.com/fonts/source-code-pro
[7]https://www.fontsquirrel.com/fonts/alegreya

```
bookdown::pdf_book:
  includes:
    in_header: latex/preamble.tex
    before_body: latex/before_body.tex
    after_body: latex/after_body.tex
  keep_tex: yes
  dev: "cairo_pdf"
  latex_engine: xelatex
  citation_package: natbib
  template: null
  pandoc_args: "--chapters"
  toc_unnumbered: no
  toc_appendix: yes
  quote_footer: ["\\VA{", "}{}"]
  highlight_bw: yes
```

All preamble settings we mentioned above are in the file `latex/preamble.tex`. In `latex/before_body.tex`, we inserted a few blank pages required by the publisher, wrote the dedication page, and specified that the front matter starts:

`\frontmatter`

Before the first chapter of the book, we inserted

`\mainmatter`

so that LaTeX knows to change the page numbering style from Roman numerals (for the front matter) to Arabic numerals (for the book body).

We printed the index in `latex/after_body.tex` (Section 2.9).

The graphical device (`dev`) for saving plots was set to `cairo_pdf` so that the fonts are embedded in plots, since the default device `pdf` does not embed fonts. Your copyeditor is likely to require you to embed all fonts used in the PDF, so that the book can be printed exactly as it looks, otherwise certain fonts may be substituted and the typeface can be unpredictable.

The `quote_footer` field was to make sure the quote footers were right-aligned: the LaTeX command `\VA{}` was provided by `krantz.cls` to include the quote footer.

The `highlight_bw` option was set to true so that the colors in syntax highlighted code blocks were converted to grayscale, since this book will be printed in black-and-white.

The book was compiled to PDF through `xelatex` to make it easier for us to use custom fonts.

All above settings except the `VF` environment and the `\VA{}` command can be applied to any other LaTeX document classes.

In case you want to work with Chapman & Hall as well, you may start with the copy of `krantz.cls` in our repository (`https://github.com/rstudio/bookdown/tree/master/inst/examples`) instead of the copy you get from your editor. We have worked with the LaTeX help desk to fix quite a few issues with this LaTeX class, so hopefully it will work well for your book if you use **bookdown**.

A

Software Tools

For those who are not familiar with software packages required for using R Markdown, we give a brief introduction to the installation and maintenance of these packages.

A.1   R and R packages

R can be downloaded and installed from any CRAN (the Comprehensive R Archive Network) mirrors, e.g., `https://cran.rstudio.com`. Please note that there will be a few new releases of R every year, and you may want to upgrade R occasionally.

To install the **bookdown** package, you can type this in R:

```
install.packages("bookdown")
```

This installs all required R packages. You can also choose to install all optional packages as well, if you do not care too much about whether these packages will actually be used to compile your book (such as **htmlwidgets**):

```
install.packages("bookdown", dependencies = TRUE)
```

If you want to test the development version of **bookdown** on GitHub, you need to install **devtools** first:

```
if (!requireNamespace("devtools")) install.packages("devtools")
devtools::install_github("rstudio/bookdown")
```

R packages are also often constantly updated on CRAN or GitHub, so you may want to update them once in a while:

```
update.packages(ask = FALSE)
```

Although it is not required, the RStudio IDE can make a lot of things much easier when you work on R-related projects. The RStudio IDE can be downloaded from https://www.rstudio.com.

A.2   Pandoc

An R Markdown document (*.Rmd) is first compiled to Markdown (*.md) through the **knitr** package, and then Markdown is compiled to other output formats (such as LaTeX or HTML) through Pandoc. This process is automated by the **rmarkdown** package. You do not need to install **knitr** or **rmarkdown** separately, because they are the required packages of **bookdown** and will be automatically installed when you install **bookdown**. However, Pandoc is not an R package, so it will not be automatically installed when you install **bookdown**. You can follow the installation instructions on the Pandoc homepage (http://pandoc.org) to install Pandoc, but if you use the RStudio IDE, you do not really need to install Pandoc separately, because RStudio includes a copy of Pandoc. The Pandoc version number can be obtained via:

```
rmarkdown::pandoc_version()
## [1] '1.17.2'
```

If you find this version too low and there are Pandoc features only in a later version, you can install the later version of Pandoc, and **rmarkdown** will call the newer version instead of its built-in version.

A.3   LaTeX

LaTeX is required only if you want to convert your book to PDF. The typical choice of the LaTeX distribution depends on your operating system. Windows users may consider MiKTeX (http://miktex.org), Mac OS X users can install MacTeX (http://www.tug.org/mactex/), and Linux users can install TeXLive (http://www.tug.org/texlive). See https://www.latex-project. org/get/ for more information about LaTeX and its installation.

Most LaTeX distributions provide a minimal/basic package and a full package. You can install the basic package if you have limited disk space and know how to install LaTeX packages later. The full package is often significantly larger in size, since it contains all LaTeX packages, and you are unlikely to run into the problem of missing packages in LaTeX.

LaTeX error messages may be obscure to beginners, but you may find solutions by searching for the error message online (you have good chances of ending up on StackExchange[1]). In fact, the LaTeX code converted from R Markdown should be safe enough and you should not frequently run into LaTeX problems unless you introduced raw LaTeX content in your Rmd documents. The most common LaTeX problem should be missing LaTeX packages, and the error may look like this:

```
! LaTeX Error: File `titling.sty' not found.

Type X to quit or <RETURN> to proceed,
or enter new name. (Default extension: sty)

Enter file name:
! Emergency stop.
<read *>

l.107 ^^M
```

---

[1]http://tex.stackexchange.com

```
pandoc: Error producing PDF
Error: pandoc document conversion failed with error 43
Execution halted
```

This means you used a package that contains `titling.sty`, but it was not installed. LaTeX package names are often the same as the `*.sty` filenames, so in this case, you can try to install the `titling` package. Both MiKTeX and MacTeX provide a graphical user interface to manage packages. You can find the MiKTeX package manager from the start menu, and MacTeX's package manager from the application "TeX Live Utility". Type the name of the package, or the filename to search for the package and install it. TeXLive may be a little trickier: if you use the pre-built TeXLive packages of your Linux distribution, you need to search in the package repository and your keywords may match other non-LaTeX packages. Personally, I find it frustrating to use the pre-built collections of packages on Linux, and much easier to install TeXLive from source, in which case you can manage packages using the `tlmgr` command. For example, you can search for `titling.sty` from the TeXLive package repository:

```
tlmgr search --global --file titling.sty
# titling:
#    texmf-dist/tex/latex/titling/titling.sty
```

Once you have figured out the package name, you can install it by:

```
tlmgr install titling  # may require sudo
```

LaTeX distributions and packages are also updated from time to time, and you may consider updating them especially when you run into LaTeX problems. You can find out the version of your LaTeX distribution by:

```
system("pdflatex --version")
## pdfTeX 3.14159265-2.6-1.40.17 (TeX Live 2016)
## kpathsea version 6.2.2
## Copyright 2016 Han The Thanh (pdfTeX) et al.
## There is NO warranty.  Redistribution of this software is
```

```
## covered by the terms of both the pdfTeX copyright and
## the Lesser GNU General Public License.
## For more information about these matters, see the file
## named COPYING and the pdfTeX source.
## Primary author of pdfTeX: Han The Thanh (pdfTeX) et al.
## Compiled with libpng 1.6.21; using libpng 1.6.21
## Compiled with zlib 1.2.8; using zlib 1.2.8
## Compiled with xpdf version 3.04
```

B

Software Usage

As mentioned in Chapter 1, this book is not a comprehensive guide to **knitr** or **rmarkdown**. In this chapter, we briefly explain some basic concepts and syntax in **knitr** and **rmarkdown**. If you have any further questions, you may post them on StackOverflow (`https://stackoverflow.com`) and tag your questions with `r`, `knitr`, `rmarkdown`, and/or `bookdown`, whichever is appropriate.

B.1 knitr

The **knitr** package was designed based on the idea of "Literate Programming" (Knuth, 1984), which allows you to intermingle program code with text in a source document. When **knitr** compiles a document, the program code (in code chunks) will be extracted and executed, and the program output will be displayed together with the original text in the output document. We have introduced the basic syntax in Section 2.3.

R Markdown is not the only source format that **knitr** supports. The basic idea can be applied to other computing and authoring languages. For example, **knitr** also supports the combination of R and LaTeX (`*.Rnw` documents), and R + HTML (`*.Rhtml`), etc. You can use other computing languages with **knitr** as well, such as C++, Python, SQL, and so on. Below is a simple example and you can see `http://rmarkdown.rstudio.com/authoring_knitr_engines.html` for more.

```
```{python}
x = 'Hello, Python World!'
```
```

```
print(x.split(' '))
```

Python users may be familiar with IPython or Jupyter Notebooks (https://jupyter.org). In fact, R Markdown can also be used as notebooks, and has some additional benefits; see this blog post for more information: https://blog.rstudio.org/2016/10/05/r-notebooks/.

If you want to show a literal chunk in your document, you can add an inline expression that generates an empty string (`` `r ''` ``) before the chunk header, and indent the code chunk by four spaces,[1] e.g.,

```
    `r ''` ``` {r}
    # a literal code chunk
    ```
```

After the document is compiled, the inline expression will disappear and you will see:

```
``` {r}
# a literal code chunk
```
```

Normally you do not need to call **knitr** functions directly when compiling a document, since **rmarkdown** will call **knitr**. If you do want to compile a source document without further converting it to other formats, you may use the knitr::knit() function.

## B.2   R Markdown

Thanks to the power of R and Pandoc, you can easily do computing in R Markdown documents, and convert them to a variety of output formats, in-

---

[1]Follow the four-space rule if the literal code chunk is to be displayed in other environments such as a list: http://pandoc.org/MANUAL.html#the-four-space-rule
```

cluding HTML/PDF/Word documents, HTML5/Beamer slides, dashboards, and websites, etc. An R Markdown document usually consists of the YAML metadata (optional) and the document body. We have introduced the syntax for writing various components of the document body in Chapter 2, and we explain more about the YAML metadata in this section.

Metadata for R Markdown can be written in the very beginning of a document, starting and ending with three dashes ---, respectively. YAML metadata typically consists of tag-value pairs separated by colons, e.g.,

```
---
title: "An R Markdown Document"
author: "Yihui Xie"
---
```

For character values, you may omit the quotes when the values do not contain special characters, but it is safer to quote them if they are expected to be character values.

Besides characters, another common type of values are logical values. Both yes and true mean true, and no/false mean false, e.g.,

```
link-citations: yes
```

Values can be vectors, and there are two ways of writing vectors. The following two ways are equivalent:

```
output: ["html_document", "word_document"]
```

```
output:
  - "html_document"
  - "word_document"
```

Values can also be lists of values. You just need to indent the values by two more spaces, e.g.,

```
output:
  bookdown::gitbook:
    split_by: "section"
    split_bib: no
```

It is a common mistake to forget to indent the values. For example, the following data

```
output:
html_document:
toc: yes
```

actually means

```
output: null
html_document: null
toc: yes
```

instead of what you probably would have expected:

```
output:
  html_document:
    toc: yes
```

The R Markdown output format is specified in the output field of the YAML metadata, and you need to consult the R help pages for the possible options, e.g., ?rmarkdown::html_document, or ?bookdown::gitbook. The meanings of most other fields in YAML can be found in the Pandoc documentation.

The **rmarkdown** package has provided these R Markdown output formats:

- beamer_presentation
- github_document
- html_document
- ioslides_presentation
- md_document
- odt_document

- `pdf_document`
- `rtf_document`
- `slidy_presentation`
- `word_document`

There are many more possible output formats in other R packages, including **bookdown**, **tufte**, **rticles**, **flexdashboard**, **revealjs**, and **rmdformats**, etc.

.

C

FAQ

Below is the *complete* list of frequently asked questions (FAQ). Yes, there is only one question here. Personally I do not like FAQs. They often mean surprises, and surprises are not good for software users.

1. Q: Will **bookdown** have the features X, Y, and Z?

    A: The short answer is no, but if you have asked yourself three times "do I really need them" and the answer is still "yes", please feel free to file a feature request to https://github.com/rstudio/bookdown/issues.

    Users asking for more features often come from the LaTeX world. If that is the case for you, the answer to this question is yes, because Pandoc's Markdown supports raw LaTeX code. Whenever you feel Markdown cannot do the job for you, you always have the option to apply some raw LaTeX code in your Markdown document. For example, you can create glossaries using the **glossaries** package, or embed a complicated LaTeX table, as long as you know the LaTeX syntax. However, please keep in mind that the LaTeX content is not portable. It will only work for LaTeX/PDF output, and will be ignored in other types of output. Depending on the request, we may port a few more LaTeX features into **bookdown** in the future, but our general philosophy is that Markdown should be kept as simple as possible.

The most challenging thing in the world is not to learn fancy technologies, but control your own wild heart.

Bibliography

Allaire, J., Cheng, J., Xie, Y., McPherson, J., Chang, W., Allen, J., Wickham, H., Atkins, A., and Hyndman, R. (2016a). *rmarkdown: Dynamic Documents for R*. R package version 1.1.

Allaire, J., R Foundation, Wickham, H., Journal of Statistical Software, Xie, Y., Vaidyanathan, R., Association for Computing Machinery, Boettiger, C., Elsevier, Broman, K., Mueller, K., Quast, B., Pruim, R., Marwick, B., Wickham, C., Keyes, O., and Yu, M. (2016b). *rticles: Article Formats for R Markdown*. R package version 0.2.

Aust, F. (2016). *citr: RStudio Add-in to Insert Markdown Citations*. R package version 0.2.0.

Chang, W. (2016). *webshot: Take Screenshots of Web Pages*. R package version 0.3.2.9000.

Cheng, J. (2016). *miniUI: Shiny UI Widgets for Small Screens*. R package version 0.1.1.

Knuth, D. E. (1984). Literate programming. *The Computer Journal*, 27(2):97–111.

R Core Team (2016). *R: A Language and Environment for Statistical Computing*. R Foundation for Statistical Computing, Vienna, Austria.

Vaidyanathan, R., Xie, Y., Allaire, J., Cheng, J., and Russell, K. (2016). *htmlwidgets: HTML Widgets for R*. R package version 0.8.

Xie, Y. (2015). *Dynamic Documents with R and knitr*. Chapman and Hall/CRC, Boca Raton, Florida, 2nd edition. ISBN 978-1498716963.

Xie, Y. (2016a). *bookdown: Authoring Books and Technical Documents with R Markdown*. R package version 0.2.

Xie, Y. (2016b). *DT: A Wrapper of the JavaScript Library 'DataTables'*. R package version 0.2.10.

Xie, Y. (2016c). *knitr: A General-Purpose Package for Dynamic Report Generation in R*. R package version 1.15.

Xie, Y. (2016d). *servr: A Simple HTTP Server to Serve Static Files or Dynamic Documents*. R package version 0.4.1.

Xie, Y. and Allaire, J. (2016). *tufte: Tufte's Styles for R Markdown Documents*. R package version 0.2.4.

Index